Springer Geography

The Springer Geography series seeks to publish a broad portfolio of scientific books, aiming at researchers, students, and everyone interested in geographical research. The series includes peer-reviewed monographs, edited volumes, textbooks, and conference proceedings. It covers the entire research area of geography including, but not limited to, Economic Geography, Physical Geography, Quantitative Geography, and Regional/Urban Planning.

More information about this series at http://www.springer.com/series/10180

Emil Vespremeanu · Mariana Golumbeanu

The Black Sea

Physical, Environmental and Historical Perspectives

Springer

Emil Vespremeanu
Faculty of Geography
University of Bucharest
Bucharest
Romania

Mariana Golumbeanu
The National Institute for Marine Research
and Development "Grigore Antipa"
Constanta
Romania

ISSN 2194-315X ISSN 2194-3168 (electronic)
Springer Geography
ISBN 978-3-319-70853-9 ISBN 978-3-319-70855-3 (eBook)
https://doi.org/10.1007/978-3-319-70855-3

Library of Congress Control Number: 2017958552

Printed on acid-free paper

This Springer imprint is published by Springer Nature
The registered company is Springer International Publishing AG
The registered company address is: Gewerbestrasse 11, 6330 Cham, Switzerland

To Prof. Dr. Fokion Vosniakos

Contents

Chapter 1
Introduction

Abstract The Black Sea is a semi-enclosed sea component of the Mediterranean Sea (European Mediterranean or Euro-African Mediterranean) to whose main basin are linked several straits and seas: Bosphorus Strait, Marmara Sea, Dardanelles Strait and Aegean Sea. Amongst the well-defined seas of Mediterranean Sea (Ionian Sea, Tyrrhenian Sea, Aegean Sea, Adriatic Sea, Alboran Sea, Azov Sea, Marmara Sea) the Black Sea has the largest area, the greatest depth, the largest volume and the greatest complexity. From a geographical point of view, the Black Sea is located in the Eastern part of South-East Europe, between 40° 55′ and 46° 32′ North latitude and 27° 27′ and 41° 42′ East longitude, in a region in which the ancient stable platforms come into contract with the young mountains generated by the alpine orogeny. The geographical location of Black Sea, as presented above, determines the main climate and biotic processes as well as the main processes of interaction between sea and atmosphere. The climate of the landlocked Black Sea can be characterized generally as continental and subject to pronounced seasonal temperature variations, although climate conditions in some parts of the basin are controlled to a great extent by the shoreline relief. A steppe climate, with cold winters and hot, dry summers, is found in the northwestern part of the basin exposed to the influence of air masses from the north. The southeastern portion of the sea, sheltered by high mountains, experiences a humid subtropical climate, with abundant precipitation, warm winters, and humid summers.

Keywords Black Sea · Mediterranean Sea · Straits · Geographical location
Climate · Temperature

1.1 Geographical Location of the Black Sea

The concept of World Ocean is quite recent and reflects the actual opinions of the scientists from Life and Earth sciences field regarding the division of Terra system in sub-systems with different ranks. According to the definition launched by the International Hydrographic Office in 1952, unanimously accepted in present, *the World Ocean* (Weltmeer) *includes all the oceans, Mediterranean seas and*

E. Vespremeanu and M. Golumbeanu, *The Black Sea*, Springer Geography,
https://doi.org/10.1007/978-3-319-70855-3_1

marginal seas from Terra. The three components are closely interlinked by energetic and substantial exchanges which ensure a dynamic integral behavior.

The oceans are basins of large sizes extended between continents, with the bottom developed on an oceanic crust and a continental crust, where significant surface, vertical and depth circulation exists, with strong interactions between aquatory and atmosphere. Three oceans are unanimously recognized: Pacific, Atlantic and Indian. The classification of Arctic basin within the category of oceans or Mediterranean seas is a controversial issue. Is discussed also the acceptance of South Ocean as an ocean unit independent from the three adjacent oceans.

The Mediterranean seas are basins with sizes smaller than the oceans, extended between the continents, separated from the adjacent ocean by means of one or several straits, with the bottom developed on a continental crust and on ocean crust, with local surface circulation, with modest interactions between sea and atmosphere. According to their size, there are large Mediterranean seas (Mediterranean Sea, American Mediterranean, Australasian Mediterranean Sea) and small Mediterranean seas (Baltic Sea, Red Sea). The Mediterranean seas consist of a variable number of basins, more or less opened towards the main basin, linked by water exchanges. From this point of view, within the Mediterranean seas open basins are identified (e.g. Alboran Sea, Tyrrhenian Sea, Adriatic Sea, Ionian Sea, of the European Mediterranean seas) and semi-enclosed seas such as, for example, Aegean Sea, Black Sea, Marmara Sea, of the same European Mediterranean.

Marginal seas are wide opened seas to the adjacent ocean, with the bottom developed especially on continental crust, many of them being shelf seas having close links with the adjacent ocean. *The semi-enclosed seas* are seas connected to the great global system which is the World Ocean by a strait or a sequence of straits and basins of small sizes (Fabbri and Fierro 1992). These are the seas from the American Mediterranean, from the Australasian Mediterranean and some seas within European Mediterranean.

The World Ocean is a dynamic and very complex global system which integrates the operation of four sub-systems: the basin with its relief developed on lithosphere, the aquatory, the biosphere and the social-economic sub-system. The World Ocean system is linked with the atmosphere together with which forms a supra-system whose structure and operation ensure the maintenance of a favorable environment for life on our planet.

According to the above definitions, *the Black Sea is a semi-enclosed sea, component of the Mediterranean Sea* (European Mediterranean or Euro-African Mediterranean) to whose main basin are linked several straits and seas: Bosphorus Strait, Marmara Sea, Dardanelles Strait and Aegean Sea (Fig. 1.1).

The Black Sea cannot be defined as continental sea due to the following:

- its basin is developed both on the continental and ocean crust;
- the basin's morphology is similar to that of ocean seas (is frequently considered a miniature ocean), with continental edges and abyssal plain;
- the aquatory has active exchange relations with the Mediterranean sea and, through it, with the rest of the World Ocean.

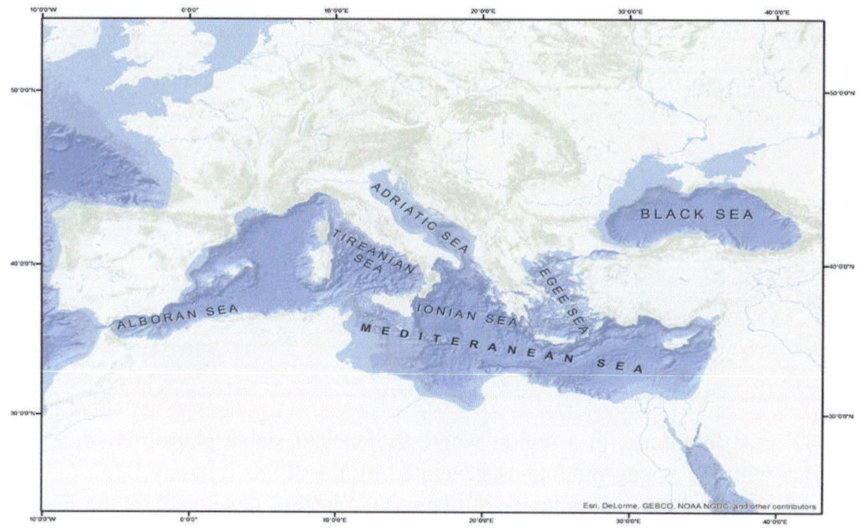

Fig. 1.1 Mediterranean Sea (European Mediterranean or Euro-African Mediterranean)

1.2 Position of the Black Sea

The Mediterranean Sea, with its component basins, including the Black Sea, represents 43% of the total area of European seas, without taking into consideration the seas of North Atlantic Ocean (Table 1.1).

The Mediterranean Sea, like all the other Mediterraneans of the World Ocean, consists of basins more or less identified within the main basin. These basins are: Black Sea, Ionian Sea, Tyrrhenian Sea, Aegean Sea, Adriatic Sea, Alboran Sea, Azov Sea, Marmara Sea (Fig. 1.1; Table 1.2). Among the well-defined seas of the Mediterranean Sea, the Black Sea has the largest area (Table 1.2).

From a geographical point of view, the Black Sea is located in the Eastern part of South-East Europe, between 40° 55′ and 46° 32′ North latitude and 27° 27′ and

Table 1.1 Position of the Mediterranean Sea between the European seas

No.	Sea	Type of sea	Area (sq. m)	%
1	Mediterranean	Large Mediterranean sea	2,971,000	43
2	Barents	Marginal sea	1,440,000	21
3	Norway Sea	Marginal sea	1,385,000	20
4	North Sea	Marginal sea	575,000	8
5	Baltic Sea	Small Mediterranean sea	414,000	6
6	White Sea	Marginal sea	90,000	1
Total			6,875,000	100

Table 1.2 Location of the Black Sea in the well-defined basins of the Mediterranean Sea

No.	Sea	Type of sea	Basin's area (sq. m)	% (of the total area)
1	Black Sea	Semi-enclosed	466,200	34.00
2	Ionian Sea	Opened	250,000	18.00
3	Tyrrhenian Sea	Opened	200,000	14.00
4	Aegean Sea	Opened	180,000	13.00
5	Adriatic Sea	Opened	130,000	9.00
6	Alboran Sea	Opened	110,000	8.00
7	Azov Sea	Semi-enclosed	38,000	3.00
8	Marmara Sea	Semi-enclosed	11,000	0.80
			1,385,000	100.00

41° 42′ East longitude, in a region where the ancient stable platforms come into contract with the young mountains generated by the alpine orogeny.

To the North there is a sequence of plains and plateaus developed on the Eastern Europe Platform and on Scythian Platform: Volhino-Podolic plateau, Doneţ Plateau, Privolgia Plateau, Black Sea Plain, which communicates, through Manici Corridor, with the Pre-Caspian Plain. *To the South* there are the Pontic Mountains with moderate altitudes (1500–2000 m), except for the Eastern part where they reach 3937 m in Kavron Peak. *To the East* the limit of Black Sea is given by the Greater Caucasus Mountains (with an altitude of maximum 5642 m in Elbrus Peak) and Small Caucasus Mountains, separated by Kolkhida Depression, by a tectonic basin recently formed following the collapse and continuous immersion of Georgian block. The Caucasus Mountains form a barrier against cold winds from North and North-East favoring the presence of a subtropical climate on the Eastern and Southern-Eastern coast of the sea. *To the West* the sea's limit is drawn on the Eastern part of low plateau developed on the Moesian and Scythian Platforms as well as on the North Dobrogean orogen.

The geographical location of the Black Sea, as presented above, determines the main climatic and biotic processes as well as the main processes of interaction between sea and atmosphere.

Latitudinal location. The fact that the middle of Black Sea is crossed by the 43° North latitude parallel, places this sea in the center of the temperate zone, which has two essential implications: the presence of well-marked seasons in accordance with the succession of solstices and equinoxes and the reception of a solar radiation which ranges between 130,000 and 150,000 cal/cm^2. This annual mean value of total radiation is lower than that received by the Mediterranean Sea, but sufficient to provide the energy necessary for the development of all the physical, chemical and biological processes.

The longitudinal location between 27° and 41° East longitude places the Black Sea in the way of air masses affected by the main baric centers: subtropical anti-cyclones of Azores, Euro-Asian continental anticyclone, North Atlantic cyclones and those from the Mediterranean Sea.

The presence of large orographic barriers—Carpathians Mountains, Balkans Mountains, Caucasus Mountains and Pontic Mountains—modifies the major circulation of atmosphere above the Black Sea imposing local climate differences. The wide opening of the Northern part of the Black Sea towards the Pontic steppes allows the entering of cold and dry air continental masses with important influences on all the processes from the Western basin. The Caucasus Mountains barrier is very efficient against the continental air masses, allowing the presence of some humid sub-tropical climate conditions in the Southern-Eastern coastal areas.

1.3 Climate of the Black Sea Region

The Black Sea is located in the temperate zone, the climate above the aquatory being decisively influenced by the main baric centers regulating the general circulation of atmosphere of South-East Europe: subtropical anticyclone of Azores and Eurasian continental anticyclone plus the cyclonic activity from North of Atlantic Ocean and that of Mediterranean Sea (Stoenescu 1960).

The climate of the Black Sea presents, on a significant part of its surface, a semi-dry character having an evaporation of 300–400 km^3/year, and an amount of rainfall of only 225–300 km^3/year (Bakan and Buyukgungor 2000).

The continental relief surrounding the Black Sea determines the very uneven distribution of the climates' dynamic factors over the aquatory. The position of Black Sea basin with wide opening, on the major part of Northern side, towards the Pontic steppes from where the dry and cold continental air masses enter, the barrier of Carpathian Mountains chain which reduces a lot the penetration of Western oceanic influences, the efficient barrier of Caucasian Mountains chain which provide protection throughout the North-East and East side against the North-East excessive continental influences, the presence, in South, of Pontic Mountains and Anatolia's Plateau determine a certain space distribution template of climate parameters.

Therefore, according to the space distribution of openings and continental barriers, we distinguish three compartments with specific climate: Western, Central and Eastern, each having a Northern and Southern sector.

The Western compartment is widely opened towards North and North-East, from where the excessive continental influences come, specific for Pontic steppes. Within this compartment, we distinguish two sectors, Northern and Southern. The Northern sector is characterized by cold and dry **winters**, with strong winds. The atmospheric precipitations vary between 38 and 50 mm. Evaporation varies between 30 and 80 mm during winter. The rainfalls have moderate value, ranging between 25 and 35 mm, and the evaporation has values between 80 and 130 mm. **Annual averages** of rainfalls vary between 350 and 600 mm, and of evaporation between 975 and 800 mm.

The East compartment is sheltered by the Caucasian Mountains which results in the specific climate with annual average temperature of 14.0–15.0 °C, with warm winters and relatively humid and moderate and humid summers (Table 1.3). This compartment is more homogenous in terms of temperature repartition but well differentiated in respect of rainfall distribution (Table 1.3) and evaporation. Therefore, in the North of Eastern compartment, the monthly average rainfalls vary between 25 and 100 mm and the evaporation between 35 and 200 mm, while in the South of this compartment the monthly average rainfalls vary between 80 and 200 mm, and the evaporation between 70 and 110 mm. The annual average of rainfalls is of 600 mm, in the North of Eastern compartment, and of 1700 mm in South. The annual average of evaporation is of 1290 mm, in North, and 785 mm in South. Locally, there are climate conditions similar to the subtropical ones, but these are due exclusively to the shelter provided by the Caucasian Mountains, as well as by the relief with South-West oriented deep valleys and sheltered depressions. Under no circumstances can we speak about a subtropical climate in Black Sea basin.

The central compartment presents transition climate features (Table 1.3).

From the above presented data the big differences between the two extreme compartments results, one under the influence of continental air masses, the other one sheltered by the Caucasian Mountains, and the transition character of the central compartment. The most significant differences are the amount of rainfalls from the North-West sector and South-East sector, which express the essential role of the barrier of Caucasian Mountains. Therefore, in the North-East corner, at 46° 28′ latitude N and 30° 10′ longitude E, the annual amount if atmospheric precipitations is of 370–380 mm, in the central part of the Black Sea, at 43° 50′ and 33° 27′, is of 340–350 mm, and in the South-East corner, at 41°55′ and 41°15′, is of 1700–1900 mm. *The wind* and aeolian processes over the sea are of high importance, determining the morphology and dynamics of waves in the mixing layer, the surface circulation, the mixing superficial processes and even the circulation through Bosphorus Strait.

The main baric centers influencing the dynamics of atmosphere above the Black Sea are (Bakan and Buyukgungor 2000):

- during summer the influence of Azores anticyclone and cyclone of Persian Bay are felt;
- during winter the center of high pressure from Asia, the center of low pressure from the Arctic regions and the baric centers of Mediterranean Sea basin are made manifest.

Under these conditions, the wind intensifications during winter have a prevailing direction from North-North-East, in the Western and central compartment, and from South and South-West in the Eastern compartment. It is noticed also a clear climate variability, the cold winters alternating with warm winters which influences to a large extent the thermal structure of the aquatory.

Table 1.3 Air temperature and rainfall amount in the space over the Black Sea (after synoptic maps of Black Sea GIS)

Month	Sector	Western compartment T (°C)	Western compartment Rainfall (mm)	Central compartment T (°C)	Central compartment Rainfall (mm)	Eastern compartment T (°C)	Eastern compartment Rainfall (mm)
I	N	−0.3 to 3	Under 40	4–5	40–60	3.5–6	80–100
I	S	4–5	60–100	5–6	60–100	6–7	100–120
VIII	N	21–22.5	25–50	22.7–22.8	25–50	22.5–22. 6	125–150
VIII	S	22.5–23	25–50	22.7–22.8	25–50	22.9–23	175–200

References

Bakan G, Buyukgungor H (2000) The Black Sea. Mar Pollut Bull 41(1–6):24–43

Fabbri P, Fierro G (ed) (1992) Semi-enclosed seas: exchange of environmental experiences between Mediterranean and Caribbean Countries. Elsevier, London, p 153

Stoenescu St (1960) Clima. Monografia Geografică a Republicii Populare Române. Vol I, Geografia Fizică, Ed Academiei Române, Bucuresti, pp 257–375

Chapter 2
Brief History of the Black Sea and Scientific Research

Abstract The Black Sea name (*Black Sea, Mer Noire, Mare Nero, Schwarze Mer, Cernoe More, Kara Deniz*) is given recently, dating since the 14th century and has no connection with the Ancient names. In the early Greek antiquity, the Black Sea was called *Pontus Axeinus* ("inhospitable sea") and later was called *Pontus Euxinos* ("hospitable sea"). The first accurate bathymetric surveys of the Black Sea were executed by R. Gotie in 1820 and 1821, the results being published in 1822. The 1970s witnessed the impressive monographs *The Black Sea. Geology, Chemistry and Biology*, edited under the coordination of Egon T. Degens and David A. Ross, as well as the results of Glomar Challenger expedition. One of the first and most interesting syntheses on Black Sea is owed to Grigore Antipa who, in 1942, published the first volume of *Black Sea monography*. The Knorr expedition of 1988 was the first one to deepen the knowledge on Black Sea based on a thorough measurement. Currently, research of the Black Sea ecosystem is being undertaken in the framework of national, regional and international projects.

Keywords *Pontus axeinus/euxinos* · Bathymetric surveys · Black Sea monography Oceanographic investigations · Research projects

2.1 Name and Origin

The Black Sea name (*Black Sea, Mer Noire, Mare Nero, Schwarze Mer, Cernoe More, Kara Deniz*) is given recently, since the beginning of 14th century, and has no connection with the Ancient name. The scientific explanation of the name's origin is still far to be resolved due to the etymologic and hydronimical difficulties, as Gh. Brătianu also emphasized: "*Amongst all the issues regarding the Black Sea, there is no other more characteristic and less known than that of name, however paradoxical such affirmation may appear*" (Brătianu 1999, p. 69).

The Indian tribes (Meotian, Sindhi and Taurine) who lived in the Northern Black Sea steppes 850–800 years before Christ called the sea *Termarum*. In the early Greek antiquity, the Black Sea was called *Pontus Axeinus* ("inhospitable sea") and

E. Vespremeanu and M. Golumbeanu, *The Black Sea*, Springer Geography,
https://doi.org/10.1007/978-3-319-70855-3_2

later was called *Pontus Euxinos* ("hospitable sea"). Other names were also used such as *Pontus Scyticus*, *Pontus Sarmaticus*, *Mare Sarmaticum* and Cimmerian Sea.

In the Middle Ages were also used several names: *Mare Majus*, *Mar Maggiore*, *Mer Maiour* (Fig. 3), *Kara Deniz* (Turkey translation for "Black Sea") and in 1338 appears for the first time its Latin name of *Mare Nigrum*, launched many years later by Dimitrie Cantemir in the scientists' world of 18th century by its works in Latin language (Dimitirie Cantemir 1710).

The controversy regarding the Black Sea name's origin, in which important personalities of the Romanian culture took part, such as B.P. Hasdeu and Gh. Brătianu, remains open. However, the explanation must not be searched in the physical characteristics of the aquatory, but, we believe, in the manner in which it was perceived by the circumeuxinic local populations according often to subjective factors.

2.2 Early History of the Black Sea

The Greek Mythology keeps the first impressions of human from the history dawn about Black Sea. From Hesiod we find that Okeanos god, the most important character of the marine Greek mythology, who mastered the Ocean, lived in a deep cave in the Black Sea bottom, from where he came out to monitor the seas. Also from the mythology we find out about the Argonauts trip on Argo ship which is considered the first expedition on sea with precise goals: finding and capturing the Golden Fleece. Most of the trip was on the Black Sea between Bosphorus and Kolhida Straits. Orpheus poet, who was on Argo ship, leaves us interesting impressions about the currents, waves and prevailing winds. The fact that the Black Sea is known since prehistory is revealed also by the legends created around Snake Island, called Leuke back then. Here existed the magic temple where Ahilleus lived, retired on the island after deification.

The first information about Black Sea we owe to Ancient philosophers, starting with Homer (1000 B.C.), Hecateu (500 B.C.), Herodot (485–425 B.C.), Aristotel (384–322 B.C.), Ptolemeu (127–151 A.C.) who left texts and cartographic drawings, some of some astoundingly precise for that time. Aristotel occupies a special place within the gallery of first scholars who tried to understand the Black Sea. Therefore, in *Meteorology*, a fundamental paper written during the second Athenian period, Aristotel describes the active circulation of waters from Azov Sea, through Black Sea, Bosphorus and Dardanelle Straits, through Aegean and Mediterranean Seas towards the Atlantic Ocean, where it enters after crossing the Hercules Columns (Gibraltar Strait). Here we find the first accurate description of currents from Bosphorus Strait as well as much other information about Black Sea and Azov Sea. It followed a long period during which Black Sea was roamed by Greek, Venetian, Turkish seamen from whom remained interesting documents regarding the configuration of shores, geographical location of harbors and sailing factors and conditions.

2.3 Scientific Research on the Black Sea

The first accurate bathymetric surveys were executed by R. Gotie in 1820 and 1821, the results being published in 1822. The expedition led by E.P. Manjanari between 1825 and 1836 aimed the bathymetric elevation of the entire Black Sea basin, the map published in 1842 indicating for the first time an overall image of bathymetry and superficial sediments.

After 1850, several Russian expeditions are carried, the substantial activity of European Commission of the Danube is started, the scientific investigations are developed in all the main harbors, the first marine investigations stations appear (Odessa 1872, Sevastopol 1879, Karadag 1902) and later the stations Kerch (1919), Novorosisk (1921), Kerson (1921), Agigea (1926), Constanta (1932), Batuni (1932), Mamaia Sat (1933), Kaliakra (1933), etc.

Amongst the most significant expeditions we consider those conducted by F. Wranghel and J.B. Spindler, between 1890 and 1894 on board of the *Cernomoret, Zaporojet* and *Donet* vessels, with the participation of researchers N. Andrusov, A. Ostrumov, I. Lebedinţev, N. Zelinski. The great merit of these expeditions is the finding of anoxic waters, rich in H_2S, the clarification of thermal and saline general structures of the aquatory, the description of the main types of sediments from the deep basin and the description of general structure of marine biome.

Grigore Antipa carries investigations on the marine fauna in 1893, for 9 months, at the board of the *Elisabeta* cruiser. It is the first Romanian oceanographic expedition in the Black Sea, in which a rich zoological material is collected and chemical measurements are carried as well as currents determinations.

Starting from 1890, the Romanian Maritime Hydrographic Directorate and the General Staff of Army perform highly accurate cartographic and oceanographic papers both in the coast area but also on the high of Black Sea. Later, the Russian expeditions conducted by N.M. Knipovici (1920–1922) and by J.M. Şokalaschi and I.A. Nikitin (1923–1926) continued the investigation of the basin's morphology and the oceanographic structure of Black Sea aquatory.

An important role regarding the information on currents goes to R. Ciocârdel who, between 1932 and 1937, arranges several research cruises on board of Romanian War Royal Marine vessel. The result of these investigations consisted in the elaboration of the most complex map of currents from the North-West of Black Sea, with genetic and dynamic oceanographic explanations underlying the recent studies. One of the first and most interesting syntheses on Black Sea is owed to Grigore Antipa who, in 1942, publishes the first volume of *Black Sea monography*. Unfortunately, the second volume remained in manuscript and was further lost.

After World War II, the oceanographic investigation of Black Sea is resumed within the specialized institutions of riparian countries, become more intense and organized until the beginning of '60s when large collaboration are established with the participation of the most important oceanographic center of the world. The preliminary investigation of Black Sea becomes now a systematic investigation.

The programs developed based on the investigations on board of *Atlantis II* (1969) and *Glomar Challenger* (1975) vessels resulted in the publication of various papers, including the impressive monographs *The Black Sea. Geology, Chemistry and Biology* edited under the coordination of the well-known oceanographers Egon T. Degens and David A. Ross, as well as the results of *Glomar Challenger* expedition (*Deep Sea Drilling Project*, vol. XVII). It was the first moment of synthesis which extends a lot the limits of Black Sea knowledge to all levels, by specifying the further approaches of deepening.

The Knorr expedition (Black Sea Oceanographic Expedition *R/V Knorr*) conducted between April 5th to August 1st, 1988, is the first one of the series of those which deepens the knowledge on Black Sea based on a stringent measurement program using the entire arsenal of modern oceanography. The scientists focused on the study of inorganic nutrients, the distribution of oxygen and hydrogen sulfide, as well as on studying the light and heat distribution within the aquatory. The Hydroblack 91 Program, conducted in 1991, aimed the study of main physical features (pressure, temperature, density) and salinity. Five vessels from Turkey, Bulgaria and Ukraine participated to this program and over 300 measurement stations in vertical profile were performed.

During 1991–1994 the international program CoMSBlack (Cooperative Marine Science Program for the Black Sea) is developed under the coordination of Prof. David Aubrey from Woods Hole Oceanographic Institution. The program was initiated by the executive board of Intergovernmental Oceanographic Commission) at the 25th UNESCO session, held in Paris in March 1992. This is the first major basin-wide international program, with 5 research vessels from Turkey, Ukraine and Bulgaria, properly fitted for the physical oceanographic researches and some chemical and biological features of the aquatory.

In 1993 the Black Sea Environmental Programme is initiated and in 1995 the International Oceanographic Committee established a territorial Committee for Black Sea for the purpose of developing a territorial GOOS programme (Global Ocean Observing System) focused mainly on the investigation of biogeochemical flows, monitoring pollution and study of eutrophication (GOOS Report 2002).

The NATO-TU Black Sea project was held during 1993 and January 1998, with the participation of over 100 researchers from Turkey and from 14 oceanographic institutions from Europe and United States of America (Workshop Report Physical and Ecological Data and Models of the Black Sea, 1998). The project aimed the development of oceanographic database for the system's future management, the development of an infrastructure for the investigations in riparian countries and the development of some interdisciplinary models of ecosystems from Black Sea biome.

During 1999–2000, the International Atomic Energy Agency (IAEA) develops a project called "The Marine Environmental Assessment of the Black Sea" aimed for the study of radioactive contamination of waters and introduction of new nuclear techniques to determine the marine pollution.

References

Brătianu G (1999) Marea Neagră. Ed Polirom. Iaşi, Romania, p 69 (in Romanian)
GOOS Report (2002) Black Sea GOOS workshop. Second Session, Poti, Georgia 22–25 May 2001. UNESCO, GOOS Report No 109. Paris

Chapter 3
Catchment Area of the Black Sea

Abstract The catchment of the Black Sea covers an area of 1,874,904 km^2 being extended in Europe and Asia Minor. Within the Black Sea catchment area four categories of catchment areas are distinguished: Western basins and North-Western basins, Crimea basins, Caucasian basins and Asia Minor basins, which transport in total 353 km^3 of water (multiannual average calculated for 1921–1988 period). The Western and North-Western basins have the largest extension, their area being of 1,520,000 km^2, which represents 82% of the total Black Sea catchment area. In this categories fall the largest Black Sea tributaries basins: Danube, Dniester, Dnieper and Bug, which transports a multiannual average liquid flow (during 1921–1988) of 261 km^3/year, representing 76% of the flow tributary to Black Sea of its entire catchment area. The Danube basin (817,000 km^2) represents 43.57% of the total area of Black Sea basin having an essential influence over its Western compartment.

Keywords Catchment area · Tributaries · Danube River · Danube Delta

3.1 General Characterizations of the Black Sea Catchment Area

The Black Sea catchment area has an area of 1,874,904 km^2 (according to *Black Sea Transboundary Diagnostic Analysis* data), being extended in Europe and Asia Minor (Fig. 3.1). Within Black Sea catchment area four categories of catchment basins are distinguished: Western basins and North-Western basins, Crimea basins, Caucasian basins and Asia Minor basins, which transport in total 353 km^3 of water (multiannual average calculated for 1921–1988 period).

The Western and North-Western basins have the largest extension, their area being of 1,520,000 km^2 which represents 82% of the total Black Sea catchment area. In this categories falls the largest Black Sea tributaries basins: Danube, Dniester, Dnieper and Bug which transport a multiannual average liquid flow (during 1921–1988) of 261 km^3/year, representing 76% of the flow tributary to Black Sea from its entire catchment area, with variations between a minimum of

E. Vespremeanu and M. Golumbeanu, *The Black Sea*, Springer Geography,
https://doi.org/10.1007/978-3-319-70855-3_3

Fig. 3.1 The catchment area of the Black Sea. *Source* http://envirogrids.net

153 km^3, in 1921, and 389 km^3, in 1970. The solid flow represented by suspended alluvia, dissolved salts, mineral nutrients and organic matter, is of roughly 150 million tonnes per year, having an essential role in the operation of aquatory system and in the distribution of recent sediments.

The Crimea basins have a total area of only 2729 km^2 representing 0.14% of the total catchment area. The largest rivers are Cernaia, Belbek, Alma and Kacea which transport a multiannual average flow (1921–1988) of 0.32 km^3 of water (0.09% of the total), with variations between a minimum of 0.07 km^3 in 1930 and a maximum of 0.67 km^3 in 1968.

The Caucasian basins have a total area of 75,000 km^2 representing 4% of the total. The most important rivers are: Cuban, Rioni, Ciorok, Inguri, Kodori, Bzib, Supsa and Mzimta. The average of multiannual liquid flows is of 43 km^3 (13% of total) with variations between a minimum of 31 km^3, in 1969, and a maximum of 57 km^3, in 1922.

The Asia Minor basins have an aggregate area of 259,550 km^2 representing 14% of the total catchment area. The most important rivers are Yesilirmak, Kâzâlirmak and Sakarya. The multiannual average flow, for the period 1930–1986, is of 36 km^3 (11% of total) with variations between 25 km^2, in 1949, and 51 km^3, in 1940.

Table 3.1 The surface of catchment areas of European seas

No.	Sea	Related catchment area (km^2)	Sea's area (km^2)	Catchment area/ sea's area ratio	Sea's volume (km^3)
1	Black Sea	2,405,000	466,200	5.15	530,000
2	Baltic Sea	1,800,00	414,000	4.34	33,000
3	Mediterranean Sea	1,300,000	2,505,000	0.51	3,754,000
4	North Atlantic Ocean	900,000			
5	North Sea	800,000	575,000	1.39	54,000
6	White Sea	700,000	90,000	7.7	4,400
7	Barents Sea	500,000	1,438,000	0.34	322,000
8	Norway Sea	200,000	1,385,000	0.14	2,408,000

3.2 Comparison with the Catchment Areas of Other European Seas

The European seas have very different catchment areas according to the relief of continent and the sizes of hydrographic network. Compared to the European seas, the catchment area of the Black Sea is the most wide and complex (Table 3.1).

The basin area/marine area ratio shows the importance of river processes in the structure and operation of aquatory (Vespremeanu 1993). Therefore, the 466,200 km^2 area of the Black Sea has a catchment area of 2,405,000 km^2, which means a basin area of 5.15 km^2 for each square kilometer of sea. The main basin of Mediterranean Sea has a totally different situation, each square kilometer having only 0.51 km^2 catchment area. The other European seas have very different values of the basin area/sea area ratio (Table 4.1). It is noted the similarity between Black Sea and Baltic Sea and the differences from others.

The North Atlantic Basin cannot be reported to its area because the value would be totally insignificant. It is noticed however the low value of related catchment areas compared to the sizes and complexity of the ocean.

3.3 Danube Basin

The Danube basin (817,000 km^2) represents 43.57% of the total area of Black Sea basin having an essential influence over its Western compartment.

The Danube, Dunai (Russian, Bulgarian, Serbian), Dona (Hungarian), Tuna (Turk), Donau (German), Danube (English), Danube (French), Danubio (Italian) is the present day name of the river which crosses the largest part of Europe from Black Forest Mountains to the Black Sea. The name, with common root, comes undoubtedly from the Latin Danubius but in time the river had several names. At the dawn of history, the Egyptian sailors called the large river "Istru". This is how the river is known in the legend about Osiris travels as well as in Orpheus poem about Argonauts trip. For a short period of time the river is called "Phison" or "Phisos" especially by the Phoenician sailors. Hesiod, Herodot, Erathostene, Strabo, Pliniu, Ptolemeu use the hydronym "Istros", widely spread during the Hellenistic period and even in the Roman one ("Ister"). The name's origin remains a controversial between the affiliation to Sanscrit, Egyptian, Scythian or Slavic. The multitude of explanations should not be surprising considering the complexity of human interferences in the Danubian space which existed even since pre-history. In the Roman antiquity two names were frequently used: Danubius, for the part of the river from the upstream of Porțile de Fier (Iron Gates) and Ister or Istros, for the lower course. The finding from Donaueschingen, in the proximity of headwaters, of an inscription dated 201 A.C., in which the name of Danubius appears (Drăghicescu 1943) seems to represent the most convincing argument attesting the Roman origin of the river's name. Other origins for this name are mentioned also: Celtic, Tracian, Galo-Roman.

Leaving aside the speculations, often of fanciful nature, we consider, together with other authors, that the actual name of river is very old and were imposed by Romans in the first centuries of our era.

The formation of Danube basin. From a structural and tectonic point of view, the Danube basin evolved on 6 fields: Alpine, Hercinic, Panonic, Carpathian, Getic and Pontic. The basin formed even since Mesozoic, in several stages, following the game of tectonic, marine and sub-aerial factors. The alpine orogeny, the last from the series of mountain generating tectonic cycles is the one that builds the Danube basin skeleton during a period of around 230 million years.

During all the stages of alpine orogeny, in the central part of Alpine-Carpathian chain, a series of depressions were maintained communicating between them: depressions of Bavaria, Moravia, Pannonic and Transylvania in communication through the strait from Iron Gates, with Gaetic, Euxinic and then with Caspian and Aral basins. All these depressions operated as marine basins until the end of Pliocene, namely until around 2.5 million years ago. As the mountains were building and rising, the hydrographic independent networks were organizing, opening towards the inland lakes and seas. At the beginning of Quaternary, the waters recede on the bottom of Pannonic and Gaetic depressions. In the freshly built-out spaces a collecting artery is forming, increasingly significant, which shall become the Danube River (Vespremeanu 1993).

The upper basin is the first one formed. The collector gathers tributaries from Alps, Black Forest Mountains, Bohemian Forest Mountains and Moravia Hills, drain the Bavarian and Wien depressions, crossing the salient from Devin, the place where the Alps junction with the Carpathians.

The middle basin is formed during the second stage. Passing through Devin gate, the Danube enters in Pannonic Depression, still occupied by the marshes remained after the backing of lake's water. The region is affected by slow tectonic submersions. The sedimentary processes are very active, leading to the deposition of enormous quantities of gravel from Carpathians and Dinaric Alps. The Danube widely bends through the center of depressions, receives new tributaries and flows through a narrow gorge through the Carpathians which shall further become the Iron Gates gorge.

The lower basin is the last one formed. When entering the Getic Depresssion, the Danube forms the Getic Sea, in its backing. The large number of tributaries coming from Carpathians forces the river to maintain as south as possible, on the contact with Pre-Balkan Platform and Dobrogea Plateau. The rivers flows through Galati gate opening towards Euxinic basin, through a large bay located in the Northern boundary of Pre-Dobrogean Depression. All these processes took place during Quaternary, namely in the last 1.5 million years.

The Holocene, last part of Quaternary (the last 10,000 years) is the period during which the river's basin is completed in its present shape. Gradually, the Black Sea level tends to stabilize itself to the actual elevation in respect of which registers slight oscillations. Therefore, it reaches a unique basic level for the entire basin which shall determine the unified evolution of Danube thalweg and of its tributaries. In the lower course the Danube Delta the Danube Flood Plain and Delta are formed. In the last 2000–2500 years, the entire basin is subject to human activities, manifested through deforestation and expansion of agriculture. Catastrophic floods appear and the solid flows increase ten times than the normal flow.

In the last 100 years, the human activities are intensified by developments (embankments, drainage, constructions of dams behind which reservoirs are formed) which completely makes artificial the entire regime of flow as well as the morphology, structure and operation of floodplains.

The Danube basin sectors. The springs of Danube from Black Forest Mountains, low and very old mountains belonging to Hercinic Mountains.

The river's place of birth is considered the confluence point of two streams, Brigach and Brege, near Dounaeschingen locality from Baden-Wurtenberg region, Germany. Here is the 2863 km which represents the route that the Danube shall go through until Black Sea, finally reaching to a basin area of 817,000 km^2. On this surface the farming lands prevail (44%) and meadows of various types and uses (30%), the forests representing only 26%. The geographic criteria allow the Danube basin to be divided into three sectors: Upper Danube, Middle Danube and Lower Danube.

Upper Danube is located between 2863 km (Donaueschingen) and 1791 km (Devin). The basin's area is of 240,000 km^2, representing 29% of the total catchment area. The altitude difference between the spring and Devin is of 360 m. The sector overlaps the Hercinic Europe and Alps Europe. Until Regensburg, the Danube receives small tributaries, both from Black Forest Mountains and Bohemia Forest Mountains, as well as from Bavaria Alps (more important is Lech river), so that it reaches here a flow of 432 m^2/s. Downstream to Passau, the Inn brings from the Alps a flow of 810 m^3/s. At Lintz, the Danube flow reaches 1060 m^3/s. When leaving the upper sector, the Danube has already an average flow of 2000 m^3/s (30.7% of the multiannual average flow at the entry in Danube Delta).

Middle Danube is located between 1791 km (Devin) and 1074 km (Baziaş). The area of this part of the basin is of 360,000 km^2 (45% of the total catchment area). The sector overlaps the Alps, Pannonic and Carpathian mountains. The altitude difference between Devin and Baziaş is of only 74 m. The main tributaries are Drava (670 m^3/s), Sava (1460 m^3/s), Morava (210 m^3/s) that come from the Austria's Alps and Dinaric Alps, as well as Tisa (814 m^3/s), from the Carpathians. Due to this substantial input, the flows increase from 2000 m^3/s to 5500 m^3/s (at Baziaş) representing 84% of the average flow at the entry in delta.

Lower Danube begins from Baziaş (1074 km) and ends at the mouths on Black Sea. The basin's area in this sector is of 217,000 km^2 representing 26% of the total catchment area. This sector overlaps the Carpathian, Balkan, Getic, Moldavian, Dobrogea Mountains, which determines the maximum geographical complexity compared to the entire basin. In this last sector are identified, by their specific geographic personality, three parts: Danube Gorge (Carpathian Danube), Getic Danube and Pontic Danube (Vespremeanu 1993).

The Danube Gorge between Baziaş and Drobeta Turnu Severin (the present day Carpathian Danube) with a length of 163 km is presented as an alternation of depression small basins and straits, each developed on a certain structure and lithology. The varied and complex morphology, the thalweg relief's energy of 53 m, the high slope of longitudinal profile with several salients, were creating, in a natural manner, one of the most picturesque landscapes in Europe. The construction of the dam from Gura Văii-Sip led to the formation of Iron Gates I Reservoir, with important consequences materialized through the modification of liquid and solid flow regime as well as to the entire landscape, as we will see further.

Main hydrological features of Lower Danube. The levels regime under natural conditions presents a polyphase behavior with repeated increases and decreases in cycles with amplitudes and durations different from one year to another. By analyzing the evolution of levels on multiannual averages results that the most frequent minimums are registered in September and October (autumn) as well as in December, January and February (winter). The most frequent maximums appear in May and June (spring time and beginning of summer).

The liquid flows regime entered in Drobeta-Turnu Severin section varies annually between a minimum of 1800 m^3/s and a maximum of 17,000 m^3/s, the multiannual average being of 5600 m^3/s. Along the longitudinal profile, the input is low: Jiu with 80 m^3/s, Olt with 215 m^3/s, Argeş with 38 m^3/s, Ialomita with 44 m^3/s, Siret with 165 m^3/s, Prut with 200 m^3/s. At the Danube Delta entry, the multiannual average flow is of 6430 m^3/s. The annual distribution of flows is closely connected to levels.

The solid flows regime varies a lot in longitudinal profile, in time and in conjunction with liquid flows. In longitudinal profile are noticed maximum values of 50–70 million tonnes/year downstream of the confluence with Jiu River and upstream of the confluence with Siret and Prut rivers. For the rest of the profile, values between 20–50 million tonnes/year are registered. Following the developments over the entire basin, the solid flows downstream of Drobeta-Turnu Severin drop at values between 15 and 40 million tonnes/year. The cause of such dramatic decreases of solid flows is connected to the accumulation of sediments in those 45 lakes formed after the hydro-technical development succeeding between Ulm and Iron Gates II, but especially in Iron Gates I Lake.

The floods regime was, in natural regime, essential for the operation of Danube Delta and Floodplain. After several and various developments which led to the pronounced artificiality of hydrological regime, the floods' role decreased a lot. According to the amplitude and duration, we differentiate floods with small amplitude and duration, with moderate amplitude and small duration, with moderate amplitude and moderate duration, with high amplitude and long duration, with very high amplitude and long duration (Vespremeanu 1993).

The winter phenomenon begins to appear in the first decade of December and cease in the last decade of March. The ice bridge is formed during cold winters starting from downstream to upstream and persists for 30–32 days under harsh conditions (winter 1953–1954, winter 1983–1984). Their melting results in the formation of thaws which results in level increases towards upstream.

Climate of Danube basin. Viewing its location, the Danube basin is located in the temperate region at the contact of three categories of major climate influences: East continental, West oceanic and South Mediterranean (CED 1966, N. Bacso 1971, A. Staneik, S. Jovanonie, 1988). *During summer* the Western circulation prevails with winds transporting moist air masses which determine the maximum of rainfalls during June. *During winter* the East and North-East circulation prevails which moves cold and dry air masses, generating a minimum of rainfalls in January and February. In *late autumn—beginning of winter,* frequently during September and December, the Mediterranean influences are felt marked by the increase of rainfalls. During summer, the Mediterranean influences manifest through warm air masses which determine the thermal maximums from July and August.

Iron Gates I Reservoir. The construction of Hydroenergetic and Navigation System from Iron Gates I is located between Gura Vaii and Sip and the reservoir formation within Danube Gorge is the most severe intervention having the most numerous and intense consequences for the artificiality of flow on Lower Danube.

The works commenced at the beginning of '60s and the water accumulation behind hydropower plant's dam took almost two years (660 days) with three different periods (Vespremeanu and Posea 1988). The firs accumulation period runs from February 10th, 1970 and April 15th, 1970, period during which the level measured at Orşova rod rises from 354 to 790 cm, therefore with 4.36 m. The second period runs from March 25th, 1971 and April 10th, 1971, when, at Orşova, the levels rise by 8 m, from 850 to 1600 cm. The third period of accumulation runs between July 20th, 1971 and November 30th, 1971 when the levels from Orşova rise by 5 m, from 1600 to 2100 cm. This last period ends the reservoir's formation process and the levels continue to oscillate according to the affluent flow (from Bazias section) and deffluent flow (exits in Dam section), between absolute levels at 62 and 68.5 m.

In the '70s and '80s, reservoir's level oscillations at Orşova have amplitudes up to 6.5 m and therefore they cause frequent changes in the reservoir's sizes and in its hydrological regime. At high retention levels (between 66 and 68.5 m), the lake's regime extends between the Dam and Svinița, the river-lake regime, between Svinița and Old Moldova, and the river regime upstream of Old Moldova. At retention levels under 65 m significant changes occur: the lake's regime is limited between the Dam and 961 km (downstream of Ogradena), the river—lake regime between Orgadena and 1005 km (between Svinița and Drencova) and the river regime upstream of 1005 km.

The most important consequences for the lower watercourse and Danube Delta are related to the serious change of solid flows regime. The Danube Gorge, which in natural regime is transited by the entire solid flow, coming from upstream, becomes a trap in which up to 90% of alluvia remain in it (Vespremeanu and Posea 1988).

The settling process is differentiated and the greatest quantity of alluvia is deposited between Drenkova (1020 km) and entry to Cazanele Mari (976 km). For example, in April 1973, a typical distribution of the quantity of suspended alluvia was registered (Table 3.2) which shows perfectly the depositing process along the reservoir (Vespremeanu and Posea 1988; Vespremeanu 1993).

Analyzing the variation of suspended alluvia quantity in Danube's longitudinal profile, it is noticed that both during dry years as well as during rainy years, the solid flows dropped seriously after 1970. It is also noticed the great difference between the alluvia's distribution in 1965 (natural regime) and 1980. The sharp drop from Orşova shows specifically the role of trap for alluvia that the Iron Gates I reservoir has. The rises of solid flow in downstream are due both to the tributaries' input as well as to the erosions from the riverbed.

The flow's artificiality involves also the modification of solid flows regime but to a lower extent and with less important consequences. We mention two directions of loss of liquid flows. The first one is connected to the completion of Main-Danube channel. Its length between Regensburg and Bamberg is of 204 km. Here, a flow of roughly 466 million m^3/year runs from the Danube's basin in Main basin. The second direction, with lower losses of flows, is due to irrigations. Only Bulgaria takes out from the Danube over 150,000 million m^3 of water per year for irrigation.

The consequences of flow's artificiality, especially of solid flow, are felt in full force not only by Danube Delta but also by the deltaic and lagoon coast as we will see further. In 2000, the population of Danube's basin was of 77,722,460 inhabitants which represent a density of 186 inhabitants/km^2. This impressive density together with the industrial and agricultural activities, contributes largely to the pollution with toxic substances, to the microbial pollution and to the release in

Table 3.2 Variation of suspended alluvia quantity (in kg/s) between Old Moldova and Iron Gates I Hydropower plant Dam (daily data) in 1973 (after Vespremeanu and Posea 1988)

Measurement date	Old Moldova	Orşova	Dam
10.04.1973	52.1	64.8	58.0
10.05.1973	400	72.0	43.4
11.06.1973	257	53.8	31.5
10.07.1973	162	33.6	22.0
12.08.1973	133	29.5	6.1
11.09.1973	111	7.2	5.4

Black Sea of some large quantities of nutrients which have a catastrophic contribution to waters' eutrophication.

The issues of nature's transformation in Danube's basin are multiple and complex. We stopped only on the most intense and those changing the flow regime. Several other issues target the urbanization and pollution with unprecedented increase.

3.4 Danube Delta Within the European Geopolitical Background

At the end of a course of over 2860 km, collecting the water from a vast hydrological basin that exceeds 8% of the area of Europe, the Danube (the second largest river of the Continent) has during the last 16,000 years built at its mouth with the Black Sea one of the most beautiful deltas in Europe, perhaps in the whole world. The Danube Delta is famous as one of the greatest wetlands of the earth and most spectacular Ramsar sites (Fig. 3.2). The wonderful natural habitats formed here offer good living conditions for an impressive number of plants and animals. Among these, reeds form one of the largest single expanses in the world, and Letea and Caraorman forests represent the northern limit for two rare species of oak that

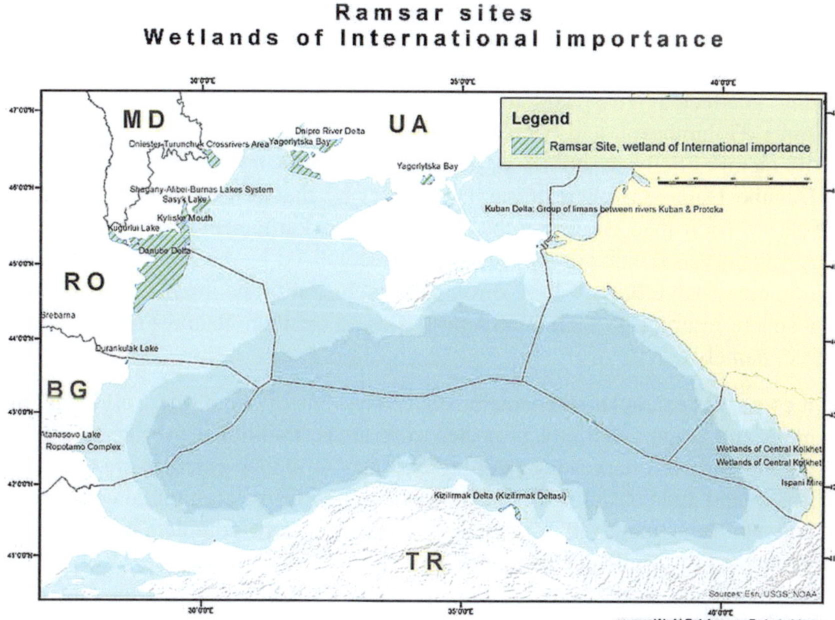

Fig. 3.2 Ramsar sites/wetlands of international importance around the Black Sea

are more frequently met in the south of the Italian and Balkan peninsulas. Together with the great number of aquatic and terrestrial plants, there are also many important colonies of pelicans and cormorants, which are characteristic of the Danube Delta, as well as a variety of other water birds which reside in or visit the delta for breeding or wintering. The large number of fish is also notable, with species of both high economic and ecological value (Golumbeanu and Nicolaev 2015).

Without doubt, the impressive range of habitats and species which occupy a relatively small area makes the Danube Delta a vital center for biodiversity in Europe, and a natural genetic bank with incalculable value for global natural heritage. Many of the plant and animal species found in the delta are also important natural resources for economic use as food, building materials and medicines, they have attracted people to the area since ancient times. The human dwellings were chiefly based on the use of these natural resources, so developing traditional economic activities and characteristic cultural and social habits. Later, there was a tendency to overexploit some of these natural resources.

This tendency, which is still seen at present time, puts increasing pressure on the resources, especially fish and grasslands, and was caused by the development of economic activities which were not in harmony with the environment; for example, sand mining at Caraorman disturbed the ecological balance, causing the loss of some areas of natural fish spawning grounds through the sedimentation and eutrophication (or nutrient enrichment) of water channels and lakes. Because of the cumulative negative effects of human activity in the delta, together with those occurring around the delta itself, there was an increasing danger that the natural ecological balance would become irreparably harmed if appropriate measures were not taken to reduce these impacts, to restore already damaged areas, to protect the existing unaffected areas, and to harness local and regional support for these measures (Golumbeanu and Nicolaev 2015).

The factors briefly described above provided arguments for the designation of the Danube Delta Biosphere Reserve (DDBR) by the Romanian Government in 1990, a decision then confirmed by the Romanian Parliament through law 82 of 1993. The universal value of the reserve was recognized by the Man and Biosphere Programme of UNESCO in 1990 through its inclusion in the international network of biosphere reserves. In fact, DDBR possesses all the main features of a biosphere reserve, namely:

(a) it conserves examples of characteristic ecosystems of one of the world's natural areas and contains strictly protected core areas, traditional use areas, e.g. for fishing and reed harvesting, and buffer zones to reduce external impacts;
(b) it is a land and coastal/marine area in which people are an integral component, and which is managed for objectives ranging from complete protection to intensive yet sustainable production;
(c) it is a regional center for monitoring, research, education and training on natural and managed ecosystems;

(d) it is a place where government decision-makers, scientists, managers and local people cooperate in developing a model programme for managing land and water to meet human needs while conserving natural processes and biological resources;
(e) it serves as a symbol of voluntary cooperation to conserve and use resources for the well-being of people everywhere.

From September 1990, the DDBR was listed as a wetland of international importance especially as waterfowl habitat under the Ramsar Convention, and is among the largest of the 600 wetlands recognized. The universal natural heritage value of the reserve was recognized in December 1990 by the inclusion of the strictly protected areas in the World Heritage List under the World Heritage Convention.

References

Drăghicescu M (1943) Istoricul principalelor puncte pe Dunare de la Gura Tisei pana la mare si pe coastele marii de la Varna la Odesa, Bucuresti, 1943, p 494 (In Romanian)
Golumbeanu M, Nicolaev S. (eds) (2015) Study on integrated coastal zone management. Ex Ponto Publishing House Constanta, original English version, ISBN 978-606-598-397-7, p 454
Vespremeanu E (1993) Oceanografie. Ed Univ. Bucureşti, p 410 (In Romanian)
Vespremeanu E, Posea A (1988) Sedimentarea şi sedimentele în lacul de acumulare Porţile de Fier. Probleme de geomorfologia României II:122 (In Romanian)

Chapter 4
Geophysics of the Black Sea Basin

Abstract The Black Sea covers an area of 466,200 km^2, while the its catchment basin's area is of 1,874,904 km^2, of which 0.817 million km^2 belong to Danube River, representing 43.57% of the total. The total length of Black Sea catchment area's watersheds is of 11,200 km, of which 8650 km on Europe territory. The Black Sea basin length between Burgas and Poti, on 45°30′ parallel is of 1200 km, while the maximum width is of 610 km and is reached in the Western part of the basin, between Oceanov and Eregli Cape. The average width is of 367.82 km. *The maximum depth* of the Black Sea is of 2245 m, according to the first Russian expeditions data. Recent measurements identified a maximum depth of only 2212 m (UNEP data). The Black Sea basin is a tectonic basin consisting of two compartments, Western and Eastern, separated by the Andrusov crest with a horst appearance. The entire region overlaps a basis developed on oceanic crust and continental crust unevenly distributed in West-East transversal profile of the basin. The relief of the Black Sea basin is divided, in terms of morphotectonic and morphostructure, in two regions: central and boundary. The central region, developed on the oceanic type basaltic crust and partially on the continental crust, encompasses the Euxinic Abyssal Plain located at depths under 2000 m. The primary tectonic relief was buried by the sediments deposited during the last 80–100 million years so that the actual relief, developed on fine and ultra-fine non consolidated sediments is particularly plain and smooth. The boundary region has a more complex structure. The relief of this region may be associated with the oceanic continental borders, showing significant differences on each of its sides. The morphogenesis of the Black Sea basin is due to the subsidence processes which followed the original continental rifting and which developed with different directions, intensities and rhythms in the two component basins of the Black Sea.

Keywords Depth · Tectonic · Ocean crust · Continental crust
Subsidence · Rifting

E. Vespremeanu and M. Golumbeanu, *The Black Sea*, Springer Geography,
https://doi.org/10.1007/978-3-319-70855-3_4

The Black Sea covers an area of 466,200 km^2, while the its catchment basin's area is of 1,874,904 km^2, of which 0.817 million km^2 belong to Danube River, representing 43.57% of the total. The total length of Black Sea catchment area's watersheds is of 11,200 km, of which 8650 km on Europe territory. The Black Sea basin length between Burgas and Poti, on 45°30' parallel is of 1200 km, while the maximum width is of 610 km and is reached in the Western part of the basin, between Oceanov and Eregli Cape. The average width is of 367.82 km.

The basin's perimeter on water line is of 4338 km, of which 1400 km on Turkey territory, 883 km in Ukraine, 992 km in Russia, 310 km in Georgia, 300 km in Bulgaria and 225 km in Romania (http://bestofukraine.com/ukraine-travel-center/crimea.html).

The maximum depth of the Black Sea is of 2245 m, according to the first Russian expeditions data. Recent measurements identified a maximum depth of only 2212 m (UNEP data). The average depth is of 1197 m. The distribution of depths is very asymmetric, fact actually reported by Murray (1900), who defines Black Sea as a deep basin with steep slopes and flat bottom. The same author establishes also the depth ratio within the basin: 35% between 0 and −35 m; only 9% between −35 and −900 m and 56% between −900 m and the greatest depths.

The recent analysis of the bathymetric map made on the basis of the last acoustic surveys (Fig. 4.1) shows a similar distribution of depths: 50% of basin's area is

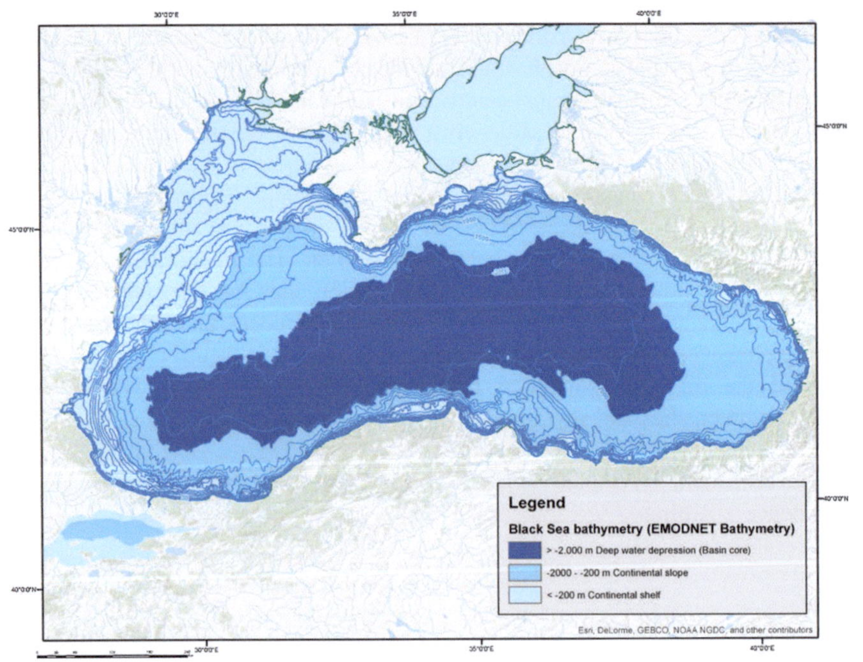

Fig. 4.1 Black Sea bathymetric map. *Source* EMODnet Bathymetry

Table 4.1 Distribution of depths in Black Sea basin

No.	Depth (m)	Area (km^2)	Ratio (%)
1	0–100	111,000	23.80
2	101–200	16,400	3.54
3	201–500	18,610	3.99
4	501–1000	31,220	6.70
5	1001–1800	51,875	11.12
6	1800–2200	237,095	50.85
	Total	466,200	100.00

covered by the abyssal plain located below −2000 m, 25% represents the area of continental shelves, and the rest of 25% is covered by continental slopes and continental glaciers (Table 4.1). This asymmetric distribution is due to basin's tectonic origin, as presented below.

The total volume of waters is of 530,000 km^3. The distribution of water volumes shows a minimum for coastal waters (50,000 km^3 or 9.5% of total) and a maximum for deep, dark, cold, oxygen less and rich in hydrogen sulphide waters (480,000 km^3, representing almost 90% of the total volume of waters). The total volume of oxygen saturated waters is of 54,000 km^3 (Sorokin 1982). The total volume of hydrogen sulphide saturated waters is of 470,000 km^3 (Sorokin 1982).

4.1 Geology and Geophysics of the Black Sea

The Black Sea basin (called also Pontic basin) is located in the Southern part of lithospheric plate close the contact with African and Arabic lithospheric plates. Geotectonic process for the formation of platforms, mountains and sedimentary basins took place here in a sequence, resulting in the current structure, tectonics and relief.

The research carried even since the second half of 19th century showed the general characteristics of Pontic Basin, bordered by young mountains in South, East and North-East, by ancient platforms and tectonic basins, heavily sedimented, in North and West. Later, the structure, extremely complicated, of the Black Sea basin was discovered as well as the tectonic and structural features of surrounding regions to which the morphogenesis of the basin is closely connected. After the World War II, the magnetic and gravitational researches and the data resulted from the seismic-acoustic analysis proved that, from a geological and geophysical point of view, the Black Sea basin develops both on the continental crust as well as on the oceanic crust.

Recent data allow for a general characterization of the geology and geophysics of the Black Sea basin (Săndulescu 1984; Kutas et al. 1998; Ionesi 1994; Nikishin et al. 2003; Cloetingh et al. 2003).

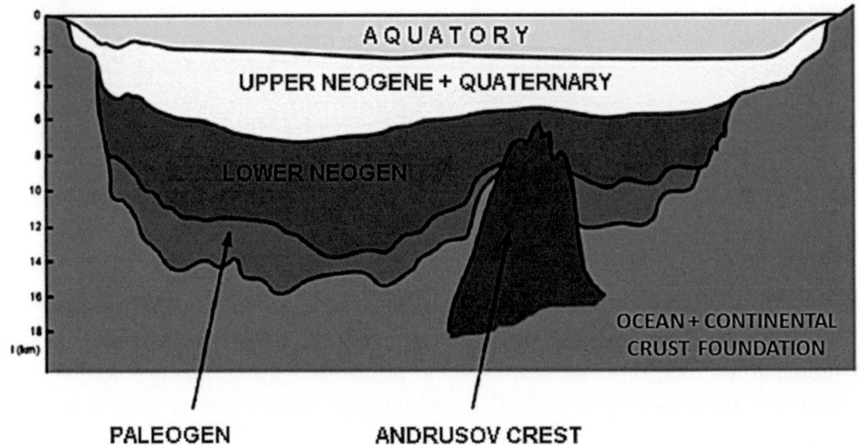

Fig. 4.2 Structure and overall tectonics of Black Sea basin on a transverse profile West-East (from Nikishin et al. 2003)

The Black Sea basin is a tectonic basin consisting of two compartments, Western and Eastern, separated by Andrusov crest with a horst appearance. The entire region overlaps a basis developed on oceanic crust and continental crust unevenly distributed in West-East transversal profile of the basin (Fig. 4.2).

The typical oceanic crust is on the central part of the Western basin, under the Pontic Abyssal Plain delimited by the isobaths of −2000 m. Here, the level of Moho discontinuity rises towards the surface, reaching in the central part of the basin only 20 km below the present day level of the sea. The basaltic layer, with a maximum thickness of 8 km, overlaps the upper mantle which has temperatures of 500–600 °C and is covered by non-consolidated sediments with a 7–14 km thickness.

The continental crust is located on the edges of Western basin and is characterized by the descending of Moho discontinuity until roughly 35 km depth, by the thickening of basaltic layer which reaches 15–18 km and by the wide extension of the granitic layer covered by other type of consolidated sediments. The Eastern basin is overlapped to a continental crust with basaltic layer of 8–9 km thickness over which is laid a thin granitic layer covered by non-consolidated sediments developed on a 10–12 km thickness. This continental crust is increasingly thickening towards East, where it reaches thicknesses of over 40 km. In the central part of the Black Sea the Andrusov Crest is located made of continental crust with thicker basaltic layer over which a thin granitic layer is laid and then the pile of non-consolidated sediments. At East of Andrusov Crest we cannot speak about typical oceanic crust due to the presence of thin granitic layer which covers the basaltic layer.

The Black Sea basin is developed on three types of different tectonic units: cratonics, pericratonics and intracratonics (Nikishin et al. 2003).

The cratonic units are represented by the Precambrian and Paleozoic crystalline platforms located in the North and East of the basin: the East-European Platform, Scythian Platform and Moesic Platform. These platforms are specific to the South-East part of the great Eurasian lithospheric plate and represented the stable boundary beyond which took place the active processes of Assyntic tectogenesis in Cambrian, Varisc in Carboniferous and Permian, Alpine in Mesosoic and Neosoic (Săndulescu 1984).

The East-European Platform, called also Eastern Europe Platform or Russian Platform, is a Precambrian crystalline plinth widened towards North and East of Black Sea basin (Săndulescu 1984).

The Scythian Platform, located in the South of East-European Platform, is younger (Lower Paleozoic) and more fractured. The contact with the East-European Platform is made through a suite of grabens on which a series of sedimentary tectonic basin developed: Bârlad Depression, Predobrogean Depression, Odessa-Sivas Depression, Indol-Kuban Depression, Manici Depression which communicated with the Peri-Caspian depressions (Săndulescu 1984).

The Moesic Platform has the same age as the Scythian Platform, but is more widened, from the bedding of Romanian Plain and Dobrogea until Rion Depression. Towards North is bounded by the North Dobrogea–Crimea–Greater Caucasus intracratonic chain and towards South is bounded by the Balkans–Istangea–Pontic–Small Caucasus pericratonic Alpine chain. The Moesic Platform was called by Stille (1953) the *Walachische Sporn* ("Walachia spur"), largely spread between the Carpathian and Caspian basin.

The intratectonic units are represented by the North-Dobrogean orogeny and Crimean orogeny, elongated until Caucasus Mountains. On the continental side of Crimea, in front of Sevastopol harbor, the Lomonosov Massif was discovered made of basalts, andesites and dacite resulted from Albian age eruptions over which lays the Upper Cretaceous and Post Cretaceous sediments (Nikishin et al. 2003).

The pericratonic units are located in the East and South of the basin representing the youngest part of the basin's boundary formed in several stages of the Alpine orogeny. This embeds the Alpine orogeny of Caucasus Mountains, Pontic Mountains and Istrangea Mountains.

The Southern border of the Black Sea basin is bordered by Srednegorie-Pontides-Achara-Trialet-Karabah Cretaceous magmatic arc (Nikishin et al. 2003).

Around the Black Sea there are a series of tectonic basins formed by continental riftogenesis in Aptian–Albian: Belgorsk basin, located in Eastern Crimea, Salgir graben from the South-East Crimea, Karkinit ditch from Odessa bay, Shtormovaya graben located in South of Karkinit ditch, Alma basin located in South-West Crimea, Kuban basin located on the North-Western flank of Greater Caucasus (Nikishin et al. 2003).

4.2 Relief of the Black Sea Basin

The relief of the Black Sea basin is divided, in terms of morphotectonic and morphostructure, in two regions: central and boundary (Brinkmann 1974; Degens and Paluska 1979; Săndulescu 1984; Vespremeanu 1989).

The central region, developed on the oceanic type basaltic crust and partially on the continental crust, encompasses the Pontic Abyssal Plain located at depths below 2000 m. The primary tectonic relief was buried by the sediments deposited during the last 80–100 million years so that the actual relief, developed on fine and ultra-fine non-consolidated sediments is particularly plain and smooth, with only of few meters of relief energy.

The boundary region has a more complex structure, as shown above. The relief of this region may be associated with the oceanic continental margins, showing significant differences on each of its sides (Vespremeanu 1989).

The Northern border may be similar to the passive continental margins, showing all the specific morphologic features: coastal plain, littoral plains, wide continental shelf, slightly steeped continental slope, wide continental glacis. The Southern and Eastern borders are similar to the active continental margins, instable, free of coastal plains, with fewer and narrowed littoral plains, with narrowed continental shelves, continental steep slopes and narrow and steep continental glacis. The Western border has mixed features closely related to structure, tectonics and morphogenesis.

The relief of Black Sea continental platform is associated with three types: Scythian, Caucasian margin and Pontic margin.

The Scythian margin, called after the broad extension of the coastal plain on the places occupied by the Scythians (Vespremeanu 1989), is located in the North basin, between Dunavăţ Promontory–Sfântu Gheorghe channel–branch from Danube Delta and Anapa Cape and is lies over the Scythian Platform and Eastern Europe Platform (Săndulescu 1984; Vespremeanu 1989). The characteristic is given by the Black Sea Plain which is a typical coastal plain, bordered towards North by the Volhino–Podolic, Doneţ and Privolgia plateaus, and towards South by several seaside plains. The coastal plain communicates towards East with the Caspian Plain through Manici corridor.

Under the Black Sea Plain there is the littoral represented by lagoon and deltaic littorals grouped in four sectors: the Danubian littoral located between Sfântu Gheorghe channel–branch (Danube Delta) and mouth and Dniester Shoreline, Fidonisic littoral, between the mouth of Dniester Firth and Tarhancut Cape, Crimea littoral, between Tarhancut Cape and Meganon Cape, Azovian littoral, between Meganon Cape and Anapa Cape.

The continental shelf shows in the North-Western sector the widest extension from the entire Black Sea basin. In the remaining Scythian margin, the continental shelf consists of a narrow notch. The continental slopes begin under the shelf's edge located at −150/−180 m, are slightly steeped, with canyons of various sizes. The greatest is Canionul Viteaz (Brave Canyon) located in the high seas of the

Romanian coast, which starts forming at the depth of −100 m and opens on the continental glacis at −1000 m through an abyssal fan.

The continental glacis are wide and are developing starting from −1500/−1800 m, supporting abyssal fans, the greater being called by some authors the Abyssal Fan of Danube (Ross et al. 1974), and by other the Pontic Abyssal Fan.

The Caucasian margin is bounded by Anapa Cape and Tsikizisi Cape, located at 25 km North of Batumi city. The main morphological feature is given by the foothill front of Caucasus Mountains continued through littoral and narrow shelves and continental steep slope until the continental glacis and Euxinic Abyssal Plain. The littoral of this margin in subdivided in two sectors: the Caucasian littoral, between Anapa Cape and mouth of Gumista river near Suhumi and Kolhide littoral between mouth of Gumists river and Cape Tsikizisi (Vespremeanu 1989).

The Pontic margin, located between Tsikizisi Cape and Koru Cape, corresponds mostly to the Pontic Mountains front which in the last 10 million years suffered a slow movement towards North that led to steep continental slope and a narrowed continental glacis. Within this margin we distinguish the Pontic littoral, between Tsikizisi Cape and Baba Cape and Bosphorus littoral, between Baba Cape and Koru Cape.

The sector between Koru Cape and Sfantu Gheorghe channel-branch (Danube Delta) has a mixed character of contact between Pontic and Scythian margins, being developed on the Moesic Platform and over the North-Dobrogean orogeny.

Islands. In the Black Sea there are three rocky islands, of continental origin, which operated as inselbergs during the regressions in ice ages: Snake Island, Berezan Island and Kefken Island. To these are added small rocky islands in Burgas Bay and various sandy barrier-islands, lagoon spits and barriers specific to the North-West littoral and Azov Sea littoral. Amongst them, Sacalin Island, located in front of Sfantu Gheorghe secondary delta, is the largest and most complex.

The overall picture of the Black Sea basin evolution according to the synthesis drafted by Brinkmann (1974), and by Săndulescu (1984) mostly confirmed by the recent detailing made by Oktay and Sakinc (1991), Kutas et al. (1998), Cloetingh et al. (2003), Nikishin et al. (2003), shows the complexity of morphogenetic processes which led to the formation of the actual basin as an inversion of relief.

On the actual location of the Black Sea basin and far West from it the Moesic Platform is formed, during the Precambrian, deformed and further elevated by the Variscan, Kimmerian and Hercinic orogeny. The high plateau thus resulted occupies a wider region far beyond the limits of the actual Black Sea basin.

The beginning of Alpine orogeny brings major changes which took place throughout its 10 phases. Therefore, in the Mesocretaceous, the Moesic Platform is affected by continental riftogenesis processes, parallel with the side formation of some small tectonic basins and some volcanic arcs preserved in the deep structure of the Black Sea actual basin, as shown above. In the Late Cretaceous begins the formation of the oceanic crust by expansion processes to the axis of the volcanic arcs and by the extension of rifts.

The opening of the Black Sea basin was created following these processes during the period between the Cenomanian and Conician (Mesocretaceous-Neocretaceous) for approximately 10 million years (Nikishin et al. 2003). Thus it begins the formation of the Black Sea basin as an inversion of relief compared to the pre-existing high plateau.

The Black Sea Western Basin, where the rifting processes started in the Upper Barremian (Lower Cretaceous or Eocretaceous) is formed the first, in the Cenomanian (Middle Cretaceous or Mesocretaceous), presenting itself as deep basin developed on oceanic crust (Cloething et al. 2003). This is proved by the sedimentary series of almost 100 million years old. The riftogenesis processes begin roughly 125 million years ago, the basin's depth rising continuously up to approx. −1600 m, depth which is reached about 35 million years ago. Afterwards, between 35 million years and 12 million years ago, the basin is filled with sediments, its depth reaching only −50 m. In the last 12 million years, by the subsidence processes, the Western basin drops again to −1200 m on the basin's boundary. In the center of the Western basin, all these processes were more intense, between 125 and 100 million years ago reaching a depth of −5000 m (Cloetingh et al. 2003).

The Black Sea Western Basin is formed later, in the Upper Paloegene, the sedimentary series being much younger (Fig. 38). The rifting processes started 60 million years ago, the subsidence being very active, but without forming a deep basin. The significant input of sediments from the Greater Caucasus and Pontic Mountains contributes to the activation of subsidence, and in the Upper Miocene the depth reached −2200 m.

The entire Black Sea basin is affected by the morphogenetic processes of Alpine orogeny that took place in Carpathian, Caucasus and Pontic Mountains, very active starting from Lower Paleogene (Nikishin et al. 2003).

Briefing the foregoing based on the last investigations owed to Hippolyte (2002), Nikishin et al. (2003) and Cloetingh et al. (2003), as well as based on some older assumptions of which the papers of Săndulescu (1984) are of main importance, we can present the following sequence of stages and morphogenetic status of the Black Sea basin:

- The stage of opening the basins through continental rifting processes developed between Upper Cretaceous and Eocene. Here we can identify a status of the formation of Western basin, between 125 and 100 million years ago, and a status of the formation of Eastern basin, between 60 and 55 million years ago.
- The stage of active subsidence developed from the moment when the basins were formed and until about 35–36 million years ago (until Eocene), when both basins are actively deepening, but with different intensity and rhythm. The tectonic subsidence rhythm is of 70–80 m/1 million years.
- The reduction stage of the subsidence intensity and massive accumulation of sediments originating from the adjoining mountain and plateau regions, developed between Oligocene and Miocene between approx. 35–36 million years ago and 5.2 million years ago, when the tectonic subsidence rhythm is of 20–30 m/1 million years.

- The completion stage of the actual basin, developed between Pliocene and Actual, in the last 5.2 million years, with the tectonic subsidence rhythm increases in both basins, the Western and Eastern, reaching up to 85 m/1 million years.

Therefore, the morphogenesis of the Black Sea basin is due to the subsidence processes which followed the original continental rifting. The series of reenactments made by Nikishin et al. (2003) clearly shows this phenomenon.

The issue of Black Sea basin subsidence is not a novelty. Such issue was approached even since the beginning of the 20th century and continued after the World War II. Stoffers and Muller (1974) affirm the assumption of a shallow basin maintained for a long period of time, after which the subsidence would have had a high velocity and took place in only 3–4 million years, between the late Pliocene and Present Day. After Degens and Paluska (1979) the shallow basin would have maintained until Middle Pleistocene, the subsidence processes becoming very active only after Cromerian (about 600,000 years ago).

The very recent investigations briefed by Nikishin et al. (2003) shows the continuous presence of the subsidence processes immediately after the original riftogenesis. Throughout those over 100 million years, the subsidence was continuous, but took place in stages, with different rhythms and intensities according to the two adjoining models. All the subsidence processes took place under a compressional intense regime which favored the heat flows in the lower part of the oceanic crust.

4.3 Bosphorus Strait

The Black Sea is connected to the Mediterranean Sea and, through it, to the World Ocean by the Turkish Straits represented by Bosphorus Strait (also called Istanbul Strait) and Dardanelles Strait (also called Canakkale Strait), which ensure an active exchange of waters, having a strong influence over the entire structure and operation of aquatory. Bosphorus Strait is located between Istanbul Peninsula (Europe) and Kocaeli Peninsula (Asia Minor), crossing a region with a very complex geological and tectonic structure (Fig. 4.2).

The research of Bosphorus Strait started since the first half of 19th century. Therefore, von Hoff (1822) is the first one dealing with the strait's morphology and issued also a morphogenetic hypothesis. The researches develop after the first serious publications, amongst which special consideration is given to Andrusov paper (1905), and at the end of 20th century it reached a good knowledge of the relief, morphogenesis and oceanography of such strait. Recently, a series of investigations carried out through state-of-art techniques contribute to the correct understanding of morphogenesis as well as of the exchange of waters between Mediterranean Sea and Black Sea both present day as well during the Quaternary period (Oguz et al. 1990; Unluata et al. 1990; Latif et al. 1991; Gokaşan et al. 1997;

Ducet et al. 1999; Bullard et al. 2000; Uchupi and Ross 2000; Gorur et al. 2001; Peneva et al. 2001; Maderich and Konstantinov 2002; Major et al. 2002; Hiskott et al. 2002; Oktay et al. 2002; Aksu et al. 2002).

The strait's length is of 31 km. The maximum depth is of 110, and the minimum depth of 33 m, the average depth 35.8 m (Yuce 1996). In the strait there are two salients, one in South, near Beşiktaş, at 33 m depth and another in North at −61 m. The strait's width ranges between a minimum of 0.7 km and a maximum of 3.5 km.

The strait is cut in the Paleozoic rocks (sandstone, clayey shales, graywake, crystalline limestones, conglomerates, arkose, quartzite), Cretacis and Miocene rocks (especially limestones and conglomerates) pierced by eruptive represented by diabasic and granite from Upper Eocene. The Bosphorus region was affected by very intense tectonic movements starting with the occurrence of North-Anatolian and Thrako–Eskisehirien faults about 3.7 million years ago and continued until now (Gokasan et al. 1997; Aksu et al. 2002; Oktay et al. 2002).

The tectonic model of Bosphorus Strait region (Oktay et al. 2002) shows the presence of two faults extended in the Northern and Southern part of the strait and several North-East-South-West oriented faults that delimit the blocks under a clockwise rotational movement.

The seismic sections shows 14 faults present in the strait's actual structure of which faults no. 13 and 14 are the most recent ones, dated to 7400–7200 years ago, which determined the collapse of old watershed from +250 m initial altitude to the present day position of −70 m (Gokosan et al. 1997). The seismic transversal profiles show the role of faults in the strait's formation (Fig. 4.3).

The morphogenetic processes and Bosphorus Strait's age were concerned by many geologists and geographers due to the importance of such issue for clarifying the relationships between Black Sea and Mediterranean Sea. Most of the authors, including also Brătescu (1942), considered that the strait has existed in the actual shape and sizes since late Pliocene, fact that was not confirmed by recent research.

The formation of Bosphorus Strait was explained by three assumptions:

- Assuming the antecedent, according to which Bosphorus Strait is a gradually deepened valley by means of fluvial erosion starting from Dacian and Levantin and at the beginning of Quaternary became a marine strait. Such assumption was launched by Von Hoff (1822), accepted and developed by Philippson (1898), Andrusov (1905), Cvijic (1908), Penk (1919), Brătescu (1942), Sholten (1974), Hsu (1978).
- Tectonic assumption according to which Bosphorus Strait was formed by a series of tectonic collapses starting about 100,000 years ago, and completed 7200 years ago, was launched by Hochstatter (1870), and recently supported by Oktay and Sakinc (1991), Ryan et al. (1997), Ryan and Pitman III (1998).
- The mixed assumption according to which the strait is the result of tectonic and fluvial processes that took place starting with late Pliocene and beginning of Quaternary (Pfannenstiel 1944; Yilmaz 2003; Gokosan et al. 1997; Aksu et al. 2002; Major et al. 2002).

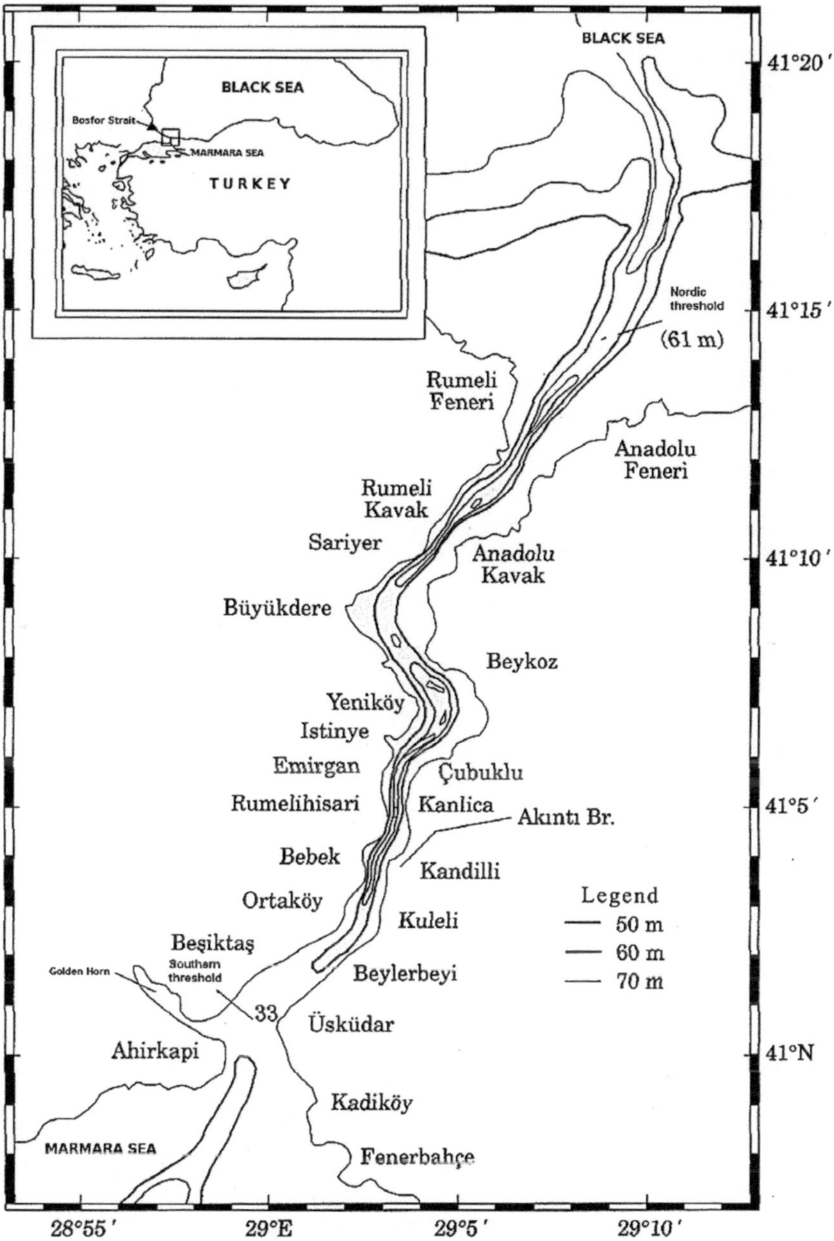

Fig. 4.3 Morphology of Bosporus Strait (*after Turkish chart TR 2912, from H. Yuce*)

Recently, Gokosan et al. (1997) conducted a study based on the use of seismic and acoustic techniques and geomorphology. The morphogenetic model developed (Fig. 12) shows the formation of the strait in four stages:

- initial stage, in which a longer Southern river was flowing in Marmara Sea and beyond the watershed, a shorter Northern river was flowing in Black Sea;
- in the second stage a tectonic basin of small sizes is formed, between Buyukdere and Beykoz, which results in a change of the hydrographic structure;
- in the third stage strong tectonic processes occur which outline the general structure of strait at South of watersheds, which is maintained towards Black Sea;
- in the fourth stage, most of the faults are reactivated, and in the North, towards the Black Sea, the watersheds collapse forming the last compartment of the strait.

The first three stages took place during Pleistocene and the fourth one after the maximum of last glaciation (Würm). The last investigations (Gorur et al. 2001; Aksu et al. 2002; Hiskott et al. 2002; Hiskott and Aksu 2002; Major et al. 2002; Mudie et al. 2002; Oktay et al., 2002) support the model drafted by Gokosan et al. (1997).

The morphogenetic model of the strait detailed above explains the formation of its relief, the circulation of waters representing another issue. The beginning of communication between Black Sea waters and Marmara Sea waters through Bosphorus Strait was dated in Quaternary (Andrusov 1905; Cvijic 1908; Penk 1919), Dacian–Levantin (Brătescu 1942); late Pleistocene–Holocene (Gokaşan et al. 1997; Ryan et al. 1997, Ryan and Pitman III 1998; Major et al. 2002; Aksu et al. 2002; Hiskott et al. 2002).

In the excellent synthesis made by Brătescu in 1942 are analyzed the previous assumptions on the opening of Bosphorus Strait, based on personal interpretations and observations. The author concludes: "*Therefore, also Bosphorus Valley began to be eroded not in the Quaternary but in Dacian and Levantin age and is older that it is supposed to be*". But, with the seriousness of a real scientist, C. Brătescu states: "*It is obvious that also our conclusions are only approximate and the new finds shall finally conclude this issue also*" (Brătescu 1942).

After the clear proof of the tectonic-erosive mechanisms included in the formation of strait, as detailed above, the communication issue between the Pontic and Marmarian basins requires the analysis of level oscillations of the two aquatories, namely defining the moment when the waters' level exceeded the Northern salient level of the strait, located at −33 m.

The communication process of Pontic waters with Marmarian waters through the strait, at the beginning of Holocene, was explained by two assumptions: (i) the continuous growth on a glacio-eustatic curve after exceeding the maximum of Würm glaciation and (ii) the catastrophic spill of Mediterranean waters which resulted in the increase of Black Sea level with about 150 m during 1 year only.

The assumption of slow glacio-eustatic growth has more versions on the contact date between the Pontic and Marmarian aquatories through Bosphorus as well as on the direction of water flow.

According to the first version, during the peak period of the last glaciation dated 20,000–18,000 years ago, the Black Sea level was at −110 m (Pirazolli 1996; Aksu et al. 2002) or at −140 m (Hiskott et al. 2002). The level is actively rising, during 10,500–10,000 years ago, reaching up to −30 m 10,000 years ago, when the Black Sea waters flows towards Marmara Sea. This flowing process is very intensive during 10,000 and 9000 years ago, and less intensive during 9000 and 6500 years ago, as resulted from the sizes and structure of the fan deposited by means of this processes on the present day shelf of Marmara Sea (Hiskott et al. 2002; Major et al. 2002). About 7200 years ago, the Black Sea level reaches to −18 m, and the warmer and saltier waters of Mediterranean Sea enter the Pontic waters through Bosphorus Strait. It is thus defined a circulation's model through strait very close to the actual one, on the surface towards Marmara Sea and on the bottom towards Black Sea.

The second version, supported by Major et al. (2002), presents two scenarios: (i) Bosphorus strait with deep salient and (ii) with artificial salient (Fig. 4.4). The sequence of episodes from the last 25,000 years reveals the evolution of exchange relationships between the three basins, Pontic, Marmarian and Aegean, as well as the directions and intensity of inputs and outputs of water in Black Sea.

The assumption of catastrophic spill of Mediterranean waters towards Black Sea was elaborated by American geologists W.B.F. Ryan and C.V. Pitman III in the '90s (Ryan et al. 1997; Ryan and Pitman III 1998). Based on the analysis of samples collected from the North of Black Sea during the Russian–American expedition of June 1993, the authors argue the following sequence of events that would have led to the actual communication through Bosphorus Strait (Fig. 4.5):

- 17,000 year ago, the Black Sea was a freshwater lake whose level was at about −140 m above the present day level.
- 14,000 years ago, the Black Sea level rose to −15 m compared to the actual one, while the level of Marmara Sea was at −60 m compared to the actual level. A fretful discharge of Black Sea waters takes place through Bosphorus towards Marmara Sea, resulting in the formation of a submerged fan on the shelf of this sea.
- 9000 years ago, the Black Sea level drops to −120 m compared to the actual one, and the Marmara Sea level rises from −60 to −30 m compared to the actual one. The Northern salient of Bosphorus is an efficient barrier maintaining the isolation of Black Sea.
- 7200 years ago, the Aegean Sea level rises to −15 m, and the Black Sea level drops to −156 m compared to the actual zero levels. The Northern salient of Bosphorus collapses resulting in the formation of a huge waterfall through which the Mediterranean waters discharge with a flow of 50–100 km^3/day (Gorur et al. 2001). This way, the Black Sea level rose from −156 to −10 m during one year only. The authors link such calamitous rise of Black Sea waters to the myth of Noah's Flood.

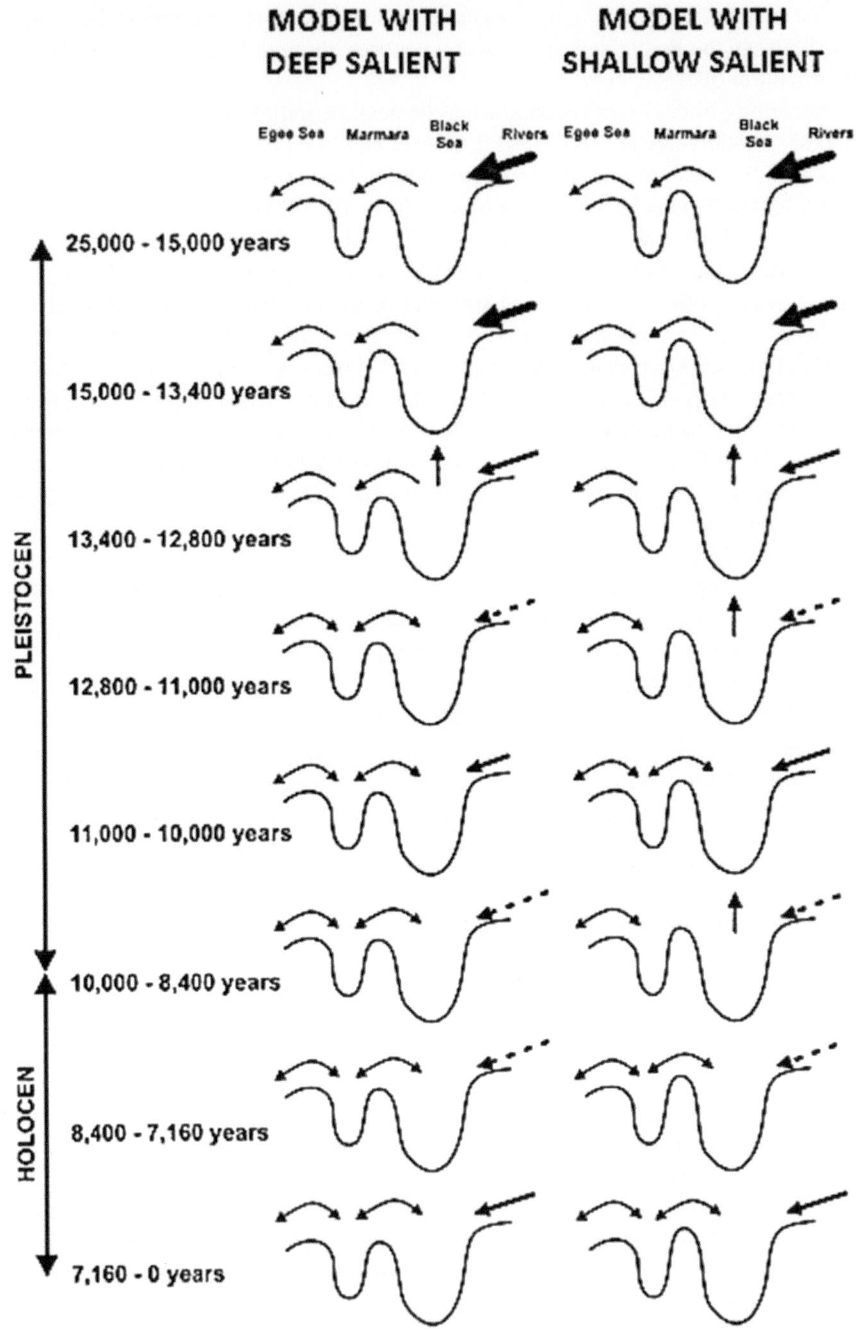

Fig. 4.4 Communication between Aegean Sea and Black Sea during Upper Pleistocene and Holocene in two assumptions: with deep salient and shallow salient (after Major et al. 2002, simplified)

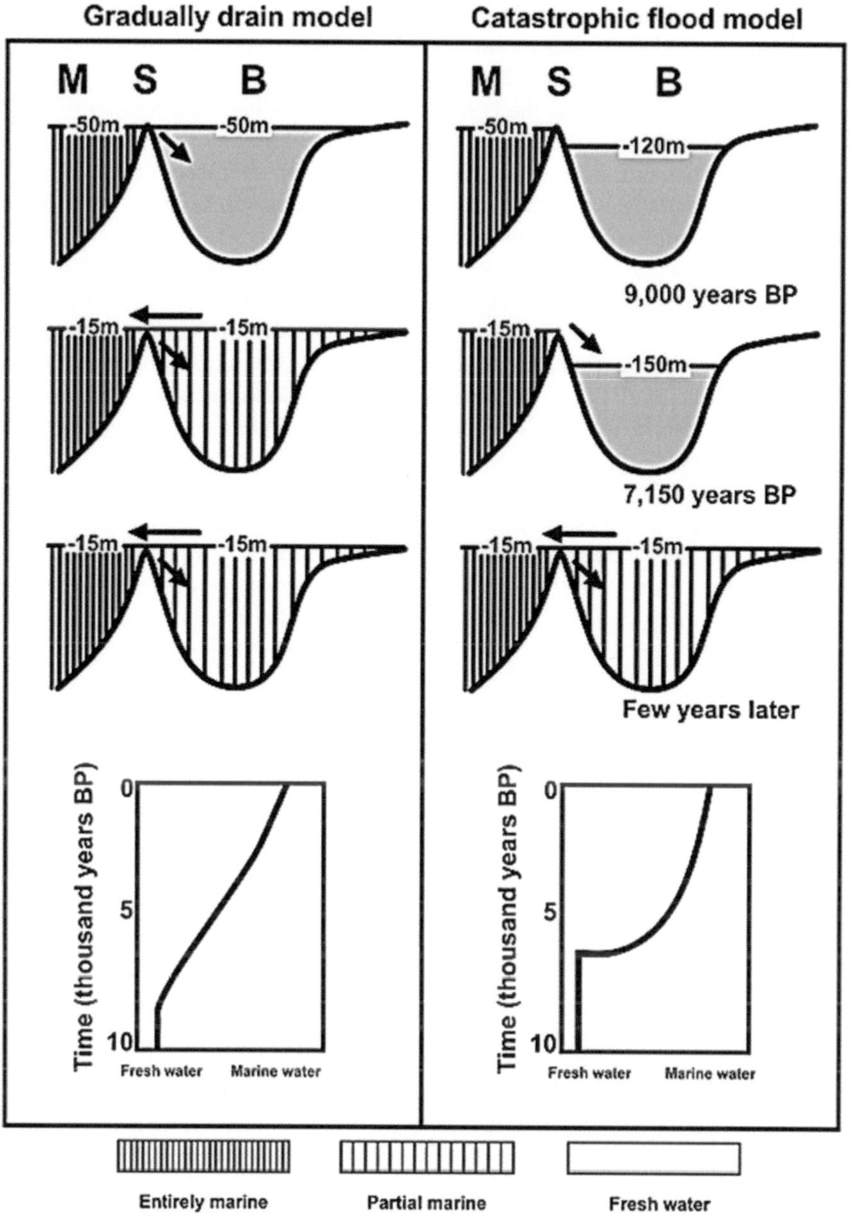

Fig. 4.5 Reconstitution of evolution of communication between Mediterranean Sea basins (M) and Black Sea (B) over Bosphorus salient (S), in two assumptions, with gradual increase of water levels and by catastrophic flood (after Ryan et al. 1997). http://www.ldeo.columbia.edu/~billr/BlackSea/Ryan_et_al_MG_1997.pdf

Widely publicized and supported by some authors (Ballard et al. 2000; Uchupi and Ross 2000), this assumption is recently fought by controversial arguments. It seems that the assumption of catastrophic discharge is closely related to reality, considering that it explains partially the isostatic behavior of Black Sea coasts following their overload with a water volume of roughly 50,000 km^3 in a very short period of time (less than a year). The exact dating of the age of sediments which form the submerged fan from Marmara Sea, as well as of sediments in front of the Northern mouth of Bosphorus strait, shall settle this dispute.

During Pleistocene, before the formation of Bosphorus Strait, the communication between the Black Sea waters and Mediterranean waters (in the broadest sense) was made through Sacarya Corridor. The assumption on this marine communication was launched by Pfannenstiel (1944) and was confirmed by very recent researches (Gorur et al. 2001). Therefore, any link between the two basins, euxinic and Mediterranean, before the end of Pleistocene must be removed from the incidence of the control imposed by Boshphorus salient.

The importance of Bosphorus Strait for Black Sea waters was first mentioned by Marsigli who performs measurements on the currents and water volumes carried by them. The general model of waters circulation through Bosphorus Strait was established even since 1885 by S.C. Makarov who carries out an in-depth study of water circulation through Bosphorus Strait based on the investigations performed during the expedition on board of Taman vessel between 1881 and 1882. The over 4000 temperature measurements and 1000 measurements of surface currents and depth currents represented the sound oceanographic knowledge of the strait. The first model of contrary water circulation and waters exchange through Bosphorus Strait was drafted.

Later, various authors, of whom we mention Andrusov (1905), Hovasse (1938), Ullyot and Ilgaz (1946), Pekash (1958), contribute to the knowledge of water circulation through Bosphorus Strait and the very recent papers bring new elements of details (Ozsoy and Beşiktepe 1997; Latif et al. 1991; Yuce 1996; Ozsoy and Unluata 1997; Peneva et al. 2001; Ivanov and Samodurov 2001; Stanev et al. 2001; Aksu et al. 2002; Maderich and Konstantinov 2002; Besiktepe 2003).

The average multiannual water flow through Bosphorus Strait is of 600 km^3/year or 20,000 m^3/s, waters with 17.90% mean salinity, from Black Sea to Marmara Sea on surface and of 300 km^3/year (10,000 m^3/s), waters with 35.50% mean salinity, from Marmara Sea to Black Sea on the bottom (Fig. 4.6).

The waters flow values through Bosphorus Strait, as well as their stratification varies a lot during the year according to the levels from Black Sea and the prevailing winds. During June and July higher levels are registered in Black Sea due to the inflow of waters from the related catchment areas. Under these conditions, the flow of Black sea waters towards Marmara Sea increases a lot and may block the access of Mediterranean waters towards Black Sea. The Northern winds have the

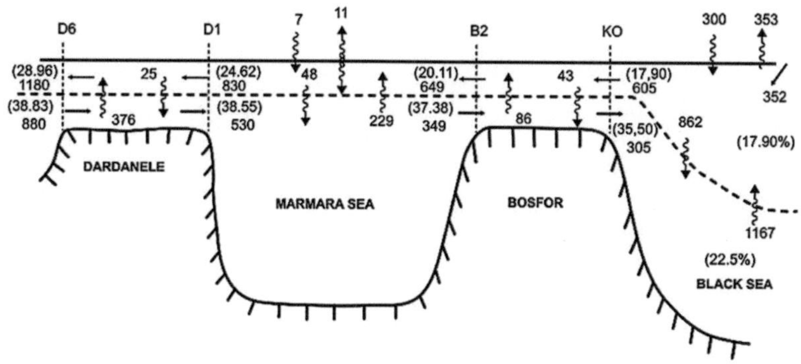

Fig. 4.6 Water exchange between Aegean Sea and Black Sea through Turkish Straits. The flows of multiannual average water flows are expressed in km^3/year (after Ozsoy and Unluata 1997, simplified)

same effect during winter, when the Black Sea waters levels drop. The wind intensifications from Southern sector may block the access of Black Sea waters in Bosphorus Strait, increasing the flow of Mediterranean waters towards North.

In the longitudinal profile of Bosphorus Strait the following stratification is present: the superficial layer in which Black Sea waters flow towards Marmara Sea, transition layer with salinity and temperature gradient and bottom layer, in which Marmara Sea waters flow towards Black Sea. Under normal conditions, free from the influences of Black Sea levels oscillations and wind intensifications, the average depths and features of this layers (Yuce 1996) are the following (Fig. 4.7):

- The superficial layer—gradient layer interface has ramps which vary in a non-linear manner: the Southern part has slight ramps which increase continuously towards North with 2 m/km, so that in the Northern end of the strait the ramp is decreasing to 0.6 m/km.
- The gradient layer—bottom layer interface has more reduced ramps, ranging between 1.4 m/km, in South, and 0.5 m/km in North.
- The gradient layer thickness is of 12 m in South, 14 m in center and 9 m in North.

This structure of Bosphorus Strait aquatory is changed should a Northern blocking (due to Black Sea high levels or prevailing winds of North) or a Southern blocking (due to Southern intense winds) is installed, when a more complicated structure is present.

The entry in Black Sea of an average quantity of 300 km^3/year, with an impressive flow of 10,000 m^3/s of Mediterranean origin water more salty and warmer than the own waters leads to, by multiples diffusions, essential exchanges of Black Sea aquatory structure, as you can see below.

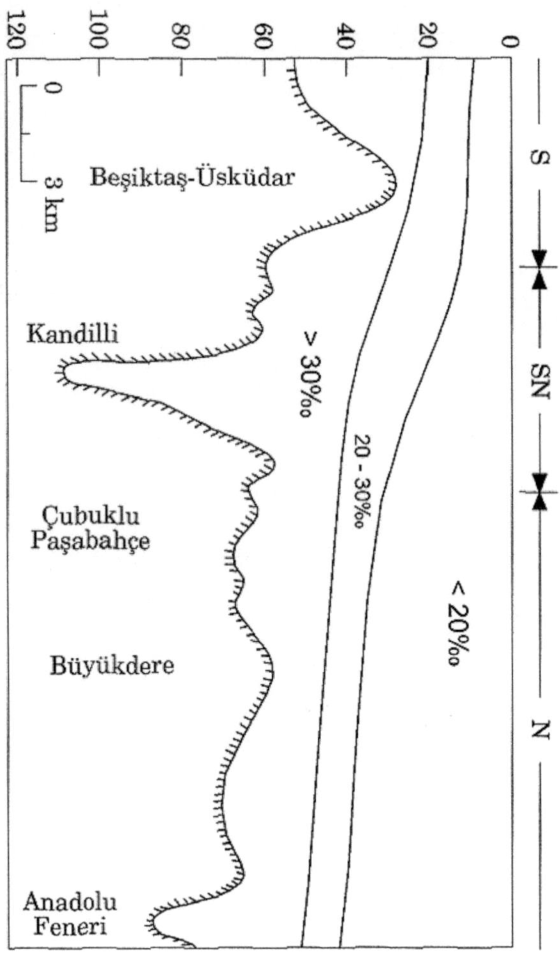

Fig. 4.7 Normal stratification of water masses, with a certain salinity, which flows through Bosphorus Strait (after Yuce 1996, amended)

4.4 The Strait of Kerch

The Strait of Kerch is less important than Bosphorus Strait, due to the small sizes of Azov Sea but especially to the drastic reduction of river input following the hydrotechnical development of Don River, completed in 1953.

The Strait of Kerch links Black Sea to Azov Sea, having a length of 45 km and a width ranging between 3.7 and 52 km. The depth varies a lot in longitudinal profile: 18 m in South, 5 m in the central part and 10 m in North. The Strait of Kerch consists of a succession of four narrow sectors, including Taman, Kerch and Kamâş Burun Bays (UNEP data).

Through the Strait of Kerch a contrary circulation takes place which ensures waters exchange with Black Sea. Therefore, from Azov Sea, between 22 and 95 km^3 of 8–10% salinity waters flows into the Black Sea, on surface, every year, and from the Black Sea between 29 and 70 km^3 of 16–18% salinity waters enters annually in Azov Sea at depths of 5–7 m (Leonov 1960).

The Azov Sea is actually a brackish firth adjoining Black Sea (the Firth of Don would be a more appropriate name) linked to it by the Strait of Kerch.

Its shape is elongated, with the big axis oriented North-East towards South-West. The maximum length is of 400 km, maximum width is of 200 km, and average width of 95 km. The Azov Sea area is of 38,800 km^2.

The maximum depth is of 14.2 m and the *average depth* of only 8.5 m. On 69% of the area there are depths between −5 and −10 m and on the remaining surface are depths between 0 and −5 m (21%) and between −10 and −14.2 m (10%). *The total volume of waters* is of 300 km^3, of which 56.7% is in the superficial layer between surface and −5 m, 36.6% between −5 and −10 m and only 6.7% below −10 m (UNEP data). The *salinity of Azov Sea waters* ranges between 7%, at River Don mouths and 11% in the proximity of the Strait of Kerch. *The average salinity* was of 10.5% in 1952 and 13.9% in 1976, such difference being caused by the influence of River Don dam, from Volgadonsk–Ţimliansk, which reduces a lot the river input (Konovalov 1995).

References

Aksu AE, Hiskott RN, Yaşar D, Işlar D, Marsh S (2002) Seismic stratigraphy of late quaternary deposits from the southwestern Black Sea shelf: evidence for non-catastrophic variations in sea-level during the last 10.000 years. Mar Geol 190:61–94

Andrusov NF (1905) Bosporus and Dardanelles: historical review of opinions of their origin. Zemlevedenie, (12)1–2:1–44 (in Russian)

Ballard RD, Coleman DF, Rosenberg GD (2000) Further evidence of abrupt holocene drowning of the Black Sea shelf. Mar Geol 170:253–261

Besiktepe S (2003) Density currents in the two-layer flow: an example of Dardanelles outflow. Oceanol Acta 26:243–253

Brătescu C (1942) Oscilaţiile de nivel ale apelor şi bazinului Mării Negre în cuaternar. Bul. Soc. Romane Regale de Geografie LXI:1–112. citările sunt din aceeaşi lucrare apărută în volumul Constantin Brătescu, Opere Alese, Ed. Ştiinţifică, 1967, pp. 193–278 (in Romanian)

Brinkmann R (1974) Geologic relation between Black Sea and Anatolia. The Black Sea Geology, Chemistry and Biology. AAPG Memoire 20:63–76

Cloetingh S, Spadini G, Van Wees JD, Beekman F (2003) Thermo-Mechanical modelling of Black Sea Basin (de)formation. Sed Geol 156:169–184

Cvijic J (1908) Grundlinien der geographie und geologie von mayzedonien und altserbien. Peterm. Mitt. Erganzngsh, vol. 162

Degens ET, Paluska A (1979) Tectonic and climatic pulses recorded in quaternary sediments of the Caspian-Black Sea region. Sed Geol 23:149–163

Ducet N, Le Traon PY, Gauzelin P (1999) Response of the Black Sea mean level to atmospheric pressure and wind forcing. J Mar Syst 22:311–327

Gokasan E, Demirbag E, Oktaz FY, Ecevitoglu B, Şimşek M, Yuce H (1997) On the origin of the Bosphorus. Mar Geol 140:183–199

Gorur N, Namic Cagatay M, Emre O, Alpar B, Sakinc M, Islam- Oglu Y, Algan O, Erkal T, Kecer M, Akkok R, Karlik G (2001) Is the abrupt drowning of the Black Sea shelf. Mar Geol 176:65–73

Hippolyte J-C (2002) Geodynamics of dobrogea (Romania): new constraints on the evolution of the Tornquist-Teisseyre Line, the Black Sea and the Carphatians. Tectonophysics 357:33–35

Hiscott RN, Aksu AE (2002) Late quaternary history of the Marmara Sea and Black Sea from high-resolution seismic and gravity-core studies. Mar Geol, 190:261–282

Hiskott RN, Aksu AE, Yaşar D, Kaminski MA, Mudie PJ, Kostylev VE, Macdonald JC, Isler FI, Lord AR (2002) Deltas south of the Bosporus strait record persistent Black Sea outflow to the Marmara Sea. Mar Geol 190:95–118. http://bestofukraine.com/ukraine-travel-center/crimea.html

Hovasse R (1938) Note oceanographique sur le bosphore. Bull. Inst. Oceanogr 739:1–8

Hsu KJ (1978) Stratigraphy of the Lacustrine Sedimentation in the Black Sea. In: Ross DA, Neprochnov Yü.P (eds) Initial reports of the deep sea drilling project, 42(2):509–524

Ionesi L (1994) Geologia unităţilor de platformă şi a orogenului Nord-Dobrogean. Ed. Tehnică, Bucureşti :280p

Ivanov LI, Samodurov AS (2001) The role of lateral fluxes in ventilation of the Black Sea. J Mar Syst 31:159–174

Konovalov SM (1995) Anthropogenic impact and ecosystems of the Black Sea. Bull de L'institut Oceanographique, Monaco 15:53–83

Kutas RI, Kobolev VP, Tsvyashchenko VA (1998) Heat flow geothermal model of the Black Sea depression. Tectonophysics 291:91–100

Latif MA, Ozosoy E, Unluata U (1991) Observations of the Mediterranean inflow into the Black Sea. Deep Sea Research 38(Supp 2):711–723

Leonov AK (1960) Reghionalnaia Okeanografia. Leningrad

Maderich V, Konstantinov S (2002) Seasonal dynamics of the system sea-straits: Black Sea-Bosphorus Case Study. Estuar Coast Shelf Sci 55:183–196

Major C, Ryan W, Lericolais G, Hajdas I (2002) Constraints on Black Sea outflow to the sea of Marmara during the last glacial-interglacial transition. Mar Geol 190:19–34

Mudie PJ, Rochon A, Aksu AE (2002) Polen stratigraphy of late quaternary cores from Marmara Sea: Land-Sea correlation and paleoclimatic history. Mar Geol 190:233–260

Murray J, Sir (1900) On the deposits of the Black Sea. Scottish Geographical Magazine December 1900

Nikishin AM, Korotaev MV, Ershov AV, Brunet MF (2003) The Black Sea Basin: tectonic history and neogene-quaternary rapid subsidence modeling. Sed Geol 156:149–168

Oguz T, Ozosoy E, Latif MA, Unluata U (1990) Modeling of hydraulically controlled exchange flow in the Bosphorus Strait. J Phys Oceanogr 20:945–965

Oktay FY, Sakinc M (1991) The Late Quaternarytectonics of the Istanbul Graben and the Origin of the Bosphorus. Terra Abstracts, Eug. 6, Strasbourg 3(1):351p

Oktay FY, Gokosan E, Sakinc M, Yaltirak C, Imren C, Demirbag E (2002) The effects of the north Anatolian fault zone on the latest connection between Black Sea and Sea of Marmara. Mar Geol 190:367–382

Ozsoy E, Beşiktepe Ş (1991) Sources of double diffusive convection and impacts on mixing in the Black Sea. In: Izdar E, Murray JM (eds) Black Sea Oceanography, Kulver

Ozsoy E, Unluata U (1997) Oceanography of the Black Sea: a review of some recent results. Earth-Sci Rev 42:231–272

Peneva E, Stanev E, Bolokopytov V, Le Traon PY (2001) Water transport in the Bosphorus Strait estimated from hydro-meteorological and altimeter data: seasonal to decade variability. J Mar Syst 31:21–33

Penk W (1919) Grundzuge der geologie des bosphorus. Veroff. Ins. Meeresk. Berlin, N.F, Heft 4:72 p

Pfannenstiel M (1944) Diluviale geologie des mittelmeergebietes, die diluvialen entwiklundgstadien und die urgeschichte von Dardanellen, Marmara Meer und Bosphorus. Geol Rundsch 34:334–342

Philippson A (1898) Bosporus Und Hellespont. Geograph. Zeitschrift, Band 4

Pirazolli PA (1996) Sea level changes: the last 20.000 years. Ed. Wiley

Ross DA, Uchupi E, Bowin CO (1974) Shallow structure of Black Sea. The Black Sea Geology, Chemistry, and Bology. AAPG Memoir 20:11–34

Ryan WBF, Pitman III W (1998) Noah's flood. The new scientific discoveries about the event that changed history. Ed. Simon and Schuster, New York

Ryan WBF, Pitman WC III, Major CO, Shimkus K, Moskalenko V, Jones JA, Dimitrov P, Gorur N, Sakinc M, Yuce H (1997) B, An abrupt drowning of the Black Sea Shelf. Mar Geol 138:119–126

Săndulescu M (1984) Geotectonica României. Ed. Tehnică, Bucureşti, 336p. (In Romanian)

Sholten R (1974) Role of the bosporus in Black Sea chemistry and sedimentation. The Black Sea geology, chemistry and biology, Aapg, Tulsa, Oklahoma

Sorokin YI (1982) The Black Sea: nature, resources. Ed. Nauka, Moskwa, 216 (In Russian)

Stanev EV, Simeonov JA, Peneva EL (2001) Ventilation of Black Sea pycnocline by the Mediterranean Plume. J Mar Syst 31:77–97

Stille H (1953) Der geotektonishe Werdegang de karpaten, Geol. Jahrb, Beih, 8:239 p

Stoffers P, Muller G (1978) Mineralogy and lithofacies of Black Sea sediments Leg. 42 B Deep Sea Drilling Project. In Black Sea–geology, chemistry and biology. AAPG Mem. 1–10, Tulsa, Oklahoma

Uchupi E, Ross DA (2000) Early Holocene marine flooding of the Black Sea. Quat Res 54:68–71

Ullyot P, Ilgaz O (1946) The hydrography of the bosporus. Geogr. Rev 36:44–66

Unluata U, Oguz T, Latif MA, Ozosoy E (1990) On the physical oceanography of the Turkish Straits. In: Pratt LJ (ed) The physical oceanography of sea straits. Edit, Kulver

Vespremeanu E (1989) Progrese în cunoaşterea reliefului litoral şi submarin cu privire specială asupra bazinului Mării Negre. Terra, 3–4 (In Romanian)

von Hoff KEA (1822) Geschichte der durch uberlieferung nachgewiessenen naturlichen veranderungen der Erdoberflache. 105–144 pp

Yilmaz Y (2003) Morphotectonic Development of the Southern Black Sea region and the surroundings. 2003 Seattle annual meeting (November 2–5, 2003), Session No. 189: "Noah Flood" and late quaternary geological and Archaeological History of the Black Sea and adjacent basins

Yuce H (1996) Mediterranean water in the strait of Istanbul (Bosphorus) and Black Sea exit. Estuar Coast Shelf Sci 43:597–616

Chapter 5
The Aquatory of the Black Sea

Abstract The issue of the evolution of Black Sea aquatory from late Neogene until present day was subject to various studies and reenactments based on the analysis and age dating of transgressive sedimentary formations. Concerning oceanographic parameters, the surface waters temperatures, as well as the air temperature are characterized by significant differences between the two compartments, Western and Eastern. The waters temperature in vertical profile presents a stratification specific for the Black Sea. In the Black Sea there are three water layers with original properties specific only for this sea. These are the cold intermediate layer (CIL), suboxic layer (SOL) and anoxic layer. To these are added the Mediterranean water masses which enter through Bosphorus Strait in the Black Sea. Recent researches proved the existence of several circulation types in the Black Sea basin: the major surface circulation from the deep basin, the sub-basin and mesoscale surface circulation, upwelling and downwelling vertical circulation. Concerning the major surface circulation in the deep basin, there is a cyclonal current called Black Sea Rim Current in the deep basin. Within the big ring of Rim Current two cyclonic vortexes are formed, Western and Eastern, with several internal vortexes with a very active dynamics, continuously changing in time and space.

Keywords Oceanographic parameters · Water masses · Rim current
Circulation

5.1 Pre-quaternary Evolution of the Black Sea Aquatory

The knowledge of the Black Sea evolution requires the analysis of two issues: the morphogenetic evolution whereby the basin reached its present day basin, briefly described above, and the evolution of the aquatory, i.e. the sequence of events which culminated in the current model of water mass in the basin. The issue of the evolution of Black Sea aquatory from late Neogene until present day was subject to various studies and reenactments based on the analysis and age dating of transgressive sedimentary formations.

E. Vespremeanu and M. Golumbeanu, *The Black Sea*, Springer Geography,
https://doi.org/10.1007/978-3-319-70855-3_5

In Upper Miocene (Messianian) the aquatory was part of the Sarmatic Sea extended from Vienna's Basin, Pannonic Basin, through the strait from Iron Gates, Getic Basin, over the current coastal plain of the Black Sea, until the Caspian and Aral basins.

At the beginning of Pliocene, the Sarmatic Sea is fragmented in smaller basins, including the Meotic Sea, which occupies the current deep basin of the Black Sea connected to the remaining World Ocean and extending more to North, over the Coastal Plain of the Black Sea and continued through the tunnel from Manici with Caspian and Aral basins. The Meotic Sea was a typical marine aquatory, with waters having a salinity of 34–35% and with prevailing biosphere of marine species.

In the Late Pliocene, the Black Sea aquatory is decreased shifting to Pontian sea-lake, with no connection with the World Ocean, maintaining only a connection with the Caspian basin. Waters were fresh, with freshwater creatures, presented in the current fauna as pontic relicts.

At the beginning of Quaternary, the Black Sea is a freshwater isolated lake, under the influence of major climate changes specific to the alternation of glacial periods with interglacial ones.

5.2 The Evolution of the Black Sea Aquatory in the Pleistocene

Starting with the later 19th century, the quaternary evolution of Black Sea aquatory was analyzed by various researchers who used sedimentological, archaeological, historic and geomorphological techniques, whose activity may be phased as following:

- The period of first researches and analysis (1880–1945) during which were laid the basis of knowledge of quaternary evolution of the Black Sea aquatory by defining their main stages and phases, as well as their nomenclature (Andrusov 1889, 1918; Arhanghelski and Strahov 1938). The first syntheses are released, amongst which the one prepared by Brătescu (1942) is the most successful.
- The period of deepening (1946–1980), during which detailed researches are carried out on sediments, fauna and chemistry, with unprecedented results (Muratov 1961; Neveskaia 1965; Degens and Ross 1970; Ross and Degens 1974; Ostrovski et al. 1977; Degens and Paluska 1979; Kuprin et al. 1980a, b; Caraivan 1981).
- The period of recent synthesis, during which, in addition to the critique analysis of previous results new arguments are brought based on the researches made by that authors (Varuscenko et al. 1984; Caraivan and Selariu 1985; Repina 1996, Ducet et al. 1999; Flood et al. 1999; Velichko 2002; Tsatskin et al. 2001; Wong et al. 2002; Yilmaz 2003; Algan 2003; Preisinger and Aslanian 2003; Cagatay 2003; Lericolais et al. 2003).

A multitude of recently briefed approaches resulted (Vespremeanu 2003). It is noted that, throughout the effort of knowledge on the Black Sea aquatory evolution, the simplified presentation of the sequence of some lake and marine basins in the favor of presenting the evolution by a sequence of transgressive and regressive stages and phases, as well as the alternation of submergences and emergences within the same stage is phased out. Also, it is noted the persistence of some numerous names for the same sedimentary sequences, as most of the authors preferred their own terms which led to a heavy terminology, with various synonymous, of ages of transgressions and regressions. In terms of geography, the oscillation of waters levels, the type of aquatory (lake or marine) as well as the connection with adjoining basins (Caspian Sea and Mediterranean Sea) is concerned.

The Quaternary history of the Black Sea aquatory is linked to the climate and tectonic changes within the basin as well as from surrounding areas. The climate conditions from Pleistocene are well known, characterized by the sequence of glacial periods with interglacial periods, which leads to the succession of transgressive phases with the regressive phases. Only in the last 40 years the climatic complexity of Holocene was established, considered for a long time as unaffected by material changes.

The climate of Holocene (the last 10,000 years) presents oscillations of lower amplitude but enough to affect, only by steric processes, the waters level, generating series of emergences and submergences on a general transgressive sequence, after R. Faibridge model (1965).

The tectonic regime is active in the central part of the basin, where the subsidence prevails, as well as on the boundaries, where elevation processes alternates with subsidence processes. Admitting the assumption of an accelerate subsidence in Middle Pleistocene (Degens and Paluska 1979), the Black Sea aquatory would have been for a long time a shallow lake from time to time connected with the Mediterranean Sea, as its deepening is caused by some recent tectonic processes.

Its connection to the Mediterranean Sea, by Bosphorus and Dardanelles straits and through Marmara and Aegean Sea was made not only according to the waters level (as C. Brătescu claimed, 1942) but also according to the morphology of the two straits, affected by intensive tectonic processes that took place on the alignment of faults and basins from the East-Anatolian space. For this reason we cannot refer the connection between the Pontic aquatory and the Mediterranean one to the actual depth of Bosphorus Strait, as assumed prior to the World War II (Brătescu 1942). It was recently proved that Bosphorus Strait is the result of recent tectonic processes, and before its opening the connection between Black Sea and Marmara Sea was made through Sakarya marine corridor (Gokasan et al. 1997; Aksu et al. 2002; Hiskott et al. 2002; Oktay et al. 2002). Besides, the communication between Black Sea and Marmara Sea through Sakarya corridor was claimed long time ago (Pfannenstiel 1944).

The connection with the Caspian Sea worked intermittently on the Manici Corridor, active since Sarmatian. According to the conclusions from the actual phase of knowledge, the evolution of the Black Sea aquatory was subject to a sequence of five stages: Ceauda, Paleo-Euxinic, Uzunlar, Karangat and

Neo-Euxinic, each having one or several transgressive-regressive stages and several submergence and emergence phases.

Further are detailed the main features of the evolution stages and phases of Black Sea aquatory in Quaternary based on the literature date from the last 100 years, briefed by us (Vespremeanu 2003), stating that the ages and limits between stages and phases are still far from being clearly defined due to the lack of sedimentary formation dating by safe techniques.

The Ceauda Stage, which took place for about 350,000 years, between 750,000 and 400,000 years B.P. (B.P., *Before Present*). During this stage, the aquatory went through two stages:

- The stage of Ceauda transgression, which took place during the Gunz-Mindel interglacial period (between 750,000 years and 500,000 years B.P.), with three phases during which the waters level was a little higher than the actual level, the Black Sea aquatory being connected to Caspian Sea and Mediterranean Sea. The fauna was formed by relict pontic species and Mediterranean species.
- The Post-Ceauda regression stage, which took place during Mindel glaciation (between 500,000 and 450,000 years B.P.), during which the waters level drops by 65–70 m below the actual level, the aquatory being a freshwater isolated lake.

The Paleo-Euxinic stage, which took place for about 125,000 years, between 450,000 years B.P. and 325,000 years B.P., with two phases:

- The Paleo-Euxinic transgression phase (also called Old Euxin) corresponding to the Mindel-Riss interglacial, between 450,000 and 375,000 years B.P., when the waters levels rise slightly above the current level, rebuilding the connection with the Mediterranean Sea.
- The Paleo-Euxinic regression stage, between 375,000 and 325,000 B.P., with amplitude of about −5 cm and a duration of only 25,000 years, on a colder climate sequence in the Mindel-Riss I interglacial.

The Uzunlar stage, with duration of about 200,000 years, between 325,000 and 125,000 years B.P. Most of the authors mentioned above agree that in this stage the following transgressive and regressive phase are succeeding:

- The Paleo-Uzunal transgression phase, between 325,000 and 320,000 years B.P., when the sea level rises over the present day levels getting in contact with the Mediterranean Sea water through Sacarya Corridor.
- The regressive phase with low amplitude, between 320,000 and 260,000 years B.P. in which the connection with the Mediterranean Sea is interrupted and returns to the isolated lake aquatory.
- The Middle Uzunlar transgression phase, between 260,000 and 240,000 years B.P. during the Riss I–Riss II interglacial, when the connection with the Mediterrranean Sea through Sakarya corridor is reinstated.
- The Middle Uzunlar regressive phase, between 240,000 and 185,000 years B.P., with the interruption of the connection with the Mediterranean Sea.

- The final Uzunlar transgression phase between 185,000 years and 170,000 years B.P., in which the sea levels was about 5 m over the actual one, the connection with the Mediterranean Sea being made through Sakarya Corridor.
- The Post-Uzunlar regression phase between 170,000 and 120,000 years B.P., during which the level of Black Sea waters was by 65–70 m below the actual one.

The Karangat stage, with duration of roughly 70,000 years, between 120,000 abd 50,000 years B.P. with two phases:

- The Karangat trasgression phase, between 120,000 and 70,000 years B.P., during the Riss II–Wurm I interglacial, when the level of waters rises until maximum 10 m over the actual levels, the contact with the Mediterranean Sea being very active through Sakarya Corridor.
- The Post-Karangat regression phase, between 70,000 and 50,000 years B.P., during the Wurm I glacial period, when the level of waters dropped until −80 m compared to the actual levels.

The Neo-Euxinic stage, with duration of about 40,000 years, which took place during the succession of glacials and interglacials specific for Wurm interval and the following postglacial period. During this stage the following phases were identified:

- The Surojski transgression stage, between 50,000 and 28,000 years B.P., during the Wurm I–Wurm II interglacial, when the waters level rose from about −80 m to about −10 m compared to the actual levels.
- The Post-Surojski regression stage between 28,000 and 20,000 years B.P., during the Wurm II and Wurm III glacial periods, when the waters level drops until −100 to 130 m.
- The Late Neo-Euxinic transgression phase, between 20,000 B.P. and Present Day. This phase corresponds to the Flandriene transgression from the Atlantic Ocean and Mediterranean Sea and overlaps the postglacial period from the late Pleistocene and Holocene. Concerning this phase numerous investigations were carried out, the results being still contradictory, especially for the lack of precise dating of the sedimentary formations and fossil remaining.

5.3 The Evolution of the Black Sea Aquatory in the Holocene

The Holocene evolution of the Black Sea aquatory, completely ignored in the past (C. Brătescu does not even recalls the possible oscillations of the sea level in Holocene), is treated today by two assumption groups:

- the first assumption considers that the Black Sea aquatory presented oscillations over a continuous transgressive curve starting from the postglacial until present day;
- the second assumption claims a catastrophic rise of the sea level.

The first assumption launched even since the beginning of the 20th century, with many contributions in the postwar period, was resumed, in the last years, based on the findings resulted from recent investigations.

The Black Sea level during the peak period of the last glaciation (about 20,000–18,000 years B.P.) was at:

- −150 m (Ryan et al. 1997; Winguth et al. 1996);
- −90 m
- −110 m (Pirazolli 1996; Aksu et al. 2002).

It was determined that the water volume between the actual level of the Black Sea and the −100 m isobath is of 54,000 km^3. Considering the volume of the actual river input (350 km^3/year) the level rise would have been made in only 155 years, without considering the evaporation losses and the intake from atmospheric precipitations (Aksu et al. 2002).

Now it is believed, almost unanimously, that the Black Sea level after the last glaciation evolved on a continuous growth curve or with several levels at which the sea level was stationing for a certain period.

Thc Russian and Ukrainian researchers claim a transgressive growth based on which there are oscillations between 2 and 15 m (Ostrovski et al. 1977).

Pirazzoli considers that the level rose starting with 17,000 year ago, on slow growths curve.

Aksu et al. (2002) proves that the sea level began to rise from −110 m, 17,000 years ago, to −100 m (15,000 years ago), reaching the actual level at only 5,000 years ago. Over this glacio-eustatic curve, the authors identify two stationing period of the sea level. Therefore, between 12,000 and 10,000 years ago the level stands at −40 m and between 9000 and 8000 years ago, the level stands at −20 m. The explanation is linked to the relations between the Black Sea and the Mediterranean Sea basins through Sacarya Corridor and Bosphorus Strait, as shown above, in the chapter where Bosphorus Strait is analyzed.

The second assumption, launched by Ryan and Pittman (1982), to which many authors have recently adhered, considers that during 9000 year B.P. and 7200 year B.P. the Black Sea was a freshwater lake whose level was by 120–156 m below the actual one (Ryan et al. 1997). At 7200 years B.P. a very fast and intense transgression took place due to the invasion of the Mediterranean Sea waters through Bosphorus Strait. The waters level rose, in a single year, from −120 m (according to some authors) or from −156 m (according to others) to about −10 m compared to the actual levels, and the authors link this natural catastrophe with the myth of Noah's flood (we described above, at Bosphorus Strait chapter, some detail issues of Ryan assumption).

The period between 7200 B.P. and present day is characterized by the general oscillation trend near the level of post-glacial transgressive logistic curve (growth curve) described above. It is about a sequence of emergences (decreasing level periods which result in the land emergence) and submergence (rising level periods which results in the land flood) in accordance with the climate evolution in Holocene (Vespremeanu 2003).

Based on recent literature data, archaeological data in Romania resulted from the researches carried out in the ancient cities of Histria, Callatis, Tomis and based on the own researches carried out on the Romanian Black Sea coast, Vespremeanu (2003) presents an attempt to identify the sequence of submergence and emergence whose features, amplitude and rhythm are for informational purposes only, to be specified in the future when the precise dating of the sedimentary formations and fossil material shall be available.

- The Pontian Emergence takes place between 7200 and 6200 years ago with a 1000 years duration, an amplitude of about 6.5 m and a rhythm of 6.5 mm/year.
- The Kalamitian Submergence takes place between 6200 and 4680 years ago with a 1520 years duration, an amplitude of 17 ma and a rhythm of about +11.18 mm/year.
- The Egresian Emergence takes place between 4680 and 4400 years ago with a 280 years duration, an amplitude of about 2.5 m and a rhythm of about −8.92 mm/year.
- The Djemetian Submergence takes place between 4400 and 4000 years ago with a 400 years duration and an amplitude of about 3 m and a rhythm of +7.5 mm/year.
- The Fanagorian Emergence takes place between 4000 and 2500 years ago, with a 1500 years duration, an amplitude of about −10 ma and a rhythm of −6.66 mm/year.
- The Hellenistic Submergence, between 2500 and 2100 years ago, with a 400 years duration, amplitude of about 7.5 m and a rhythm of +18.75 mm/year.
- The Roman Emergence, between 2100 and 1800 years ago with a 300 years duration, amplitude of 1.3 m and rhythm of −4.33 mm/year.
- The Medieval Submergence takes place between 1800 and 800 years ago, with a 1000 years duration, 3 m amplitude and a rhythm of +3 mm/year.
- The Kersonian Emergence takes place between 800 and 200 years ago, with a 600 years duration and −1 m amplitude and a rhythm of 0.6 mm/year.
- The present day Submergence, from the last 200 years but visible only after 1950, as shown below.

We mention the provisional character of the sea waters elevation mark and of the velocity of their rising and decrease processes.

Now it is very difficult to precisely determine the intensity and rhythm of oscillation processes as they are the result of the convergence of at least three factors: the development of global eustatism sent to the Black Sea, the neo-tectonic processes affecting the coastal areas in a different manner (in a positive, negative or stationary manner), climate changes.

The real curve of sea level evolution in Holocene could be precisely defined only when sufficient precise dating shall be available to us concerning the age of transgressive sedimentary formations and the horizons of marshes from the deltaic and lagoon littoral plains representing the archive in which all the sea level changes traces are preserved.

5.4 Oscillations of the Black Sea Level

The issue of Black Sea level rise appeared after 1940 when positive trend are marked on the Russian and Romanian coast (Cotovu 1958). Further, especially in Romania, the issue of sea level oscillations concerns a series of authors who have very interesting results (Banu 1961; Bondar and Filip 1962; Bondar 1974; Vespremeanu 1986, 1987, 2003; Malciu and Diaconu 2000; Dachev 2000).

Multiannual oscillations. The analysis made on series for 30 up to 130 years shows rises of the sea level with a rhythm of 5.66 mm/year at Odessa, 2.02 mm/year at Nikolaev, 0.84 mm/year at Sevastopol, 2.25 mm/year at Novorossiysk, 1.29 mm/year at Tuapse, 0.83 mm at Batumi, 5.94 mm/year at Poti (Black Sea Environmental Programme), 2.386 mm/year at Varna (Dachev 2000).

It is obvious that we are witnessing rises of the Black Sea level with rhythms between 0.83 mm/year and 5.66 mm/year, according to the various compartments of the coast. The very high differences may be explained only by the different isostasic behavior of continental margins. Such processes took place also in the past, giving an explanation for the emergent and submergent oscillations with quite large amplitude of the sea level obviously indicated by archaeological data (Vespremeanu 2003). For knowledge on the Black Sea level oscillation on the Romanian coast we have available the data recorded at the level gauges from Sulina (1856–2002) and Constanta (1933–2001).

The multilevel data series recorded at Constanta tide gauge starts in 1933 and was processed by different authors during 1933–2000. After processing a 22 years data row, Banu (1961) reached the conclusion that the growth trend of sea level is of 4.250 mm/year. Selariu, on a 36 years series indicates a 2.56 mm/year trend, Leabu, on 53 years series indicates a growth trend of 2.27 mm/year, Vespremeanu, on a 59 years series indicates a trend of 2.15 mm/year, Malciu and Diaconu (2000), on a 65 years series—1.28 mm/year, and Vespremeanu (2003), on a 70 years series draws the conclusion that the actual rise rhythm of the sea level at Constanta is of 1.30 mm/year. The variation speed of crustal movements at Constanta is of −0.008 mm/year and does not affect the sea level value.

Table 5.1 Deviations of average decennial level from the multiannual average level at Constanta (after Vespremeanu 2003)

No.	Period	Average (cm)	Deviation from the multiannual average (cm)
1.	1933–1942	11.260	−3.09
2.	1943–1952	7.580	−6.77
3.	1953–1962	15.811	−1.46
4.	1963–1972	19.535	+5.18
5.	1973–1982	16.724	+2.37
6.	1983–1992	14.939	+0.60
7.	1993–2001	18.100	+3.75

Table 5.2 Deviations of Black Sea level at Constanta from the multiannual average (1933–2001) in two characteristic periods (after Vespremeanu 2003)

No.	Period	Average (cm)	Deviation from the multiannual average (cm)
1.	1933–1962	11.553	−2.79
2.	1963–2001	17.64	+3.29

The multiannual average of Black Sea level al Constanta for 1933–2001 is of 14.350 cm. The decennial averages are negatively deviating from this multiannual average value during 1933–1962, and positively during 1963–2001 (Table 5.1).

These tendencies confirm those from Sulina and are within the global average values of the World Ocean level rise of 1–2 mm/year following global warming and related complex phenomena (Vespremeanu 2003) (Table 5.2).

The processing of the two data rows on Black Sea levels in the last 150 years shows three different periods:

- period with oscillations below the multiannual average between 1856 and 1910;
- period with oscillations around the multiannual average between 1911 and 1960;
- period with accelerated growth, after 1961.

Annual oscillations. The processing of level oscillation during a year (Vespremeanu 2003; Malciu and Diaconu 2000) shows significant seasonal differences:

- during winter, moderate levels are recorded but with large amplitude oscillations, especially in positive field;
- during spring the highest sea levels are recorded, but the oscillations amplitude is moderate and develops only in positive field;
- during summer the sea levels are low with higher oscillations in positive field;
- during autumn the lowest sea levels values are recorded, with oscillation in both fields.

All these oscillations are closely correlated with the variation of Danube flows at the river mouths (Bondar 1972; Malciu and Diaconu 2000; Vespremeanu 2003).

Daily oscillations. In 24 h are recorded oscillation caused by tides and wind intensifications. In the Western Black Sea there are mixed tides, with two maximums and two minimums in 24 h, to which secondary oscillations are added and the maximum amplitude does not exceed 0.20 m. This microtidal regime is visible only on beach faces gently sloping under 0.5° by changing of water line with 4–6 m. Winds with intensity over 10 m/s generate changes in sea level with different values according to their speed and direction. The highest sea levels rises are recorded on a wind circulation from North-East, East and South-East. The winds from South-Western, Western and North-Western sector generate decreases of the sea level.

The correlations between sea level oscillations and atmospheric pressure oscillations are insignificant due to the specific of Black Sea and circulation through Bosphorus (Ducet et al. 1999).

5.5 Hydric Balance

The hydric balance of the Black Sea concerned the oceanographists even since the late 19th century and was resumed by 33 authors in 27 papers released in 20th century. The difficulties of quantitative estimations of input and output components of the balance, especially on the level of atmospheric precipitations and evaporations, determined the existence of some very different quantitative data of the hydric balance parameters.

The equation of hydric balance in the form that takes into account inputs and outputs is:

$$Q + P + K + B = E + K_1 + B_1,$$

where: Q = river inputs; P = atmospheric precipitations inputs; K = inputs through Bosphorus Strait; B = inputs through Kerch Strait from Azov Sea; E = evaporation outputs; K_1 = outputs through Bosphorus Strait; B_1 = outputs through Kerch Strait towards Azov Sea.

The river inputs represents the component that may be calculated with more accuracy compared to others, due to the several hydrological stations existing on most of the rivers tributaries of the Black Sea as shown above (*Black Sea GIS*, 1998) (Table 5.3).

The inputs from atmospheric precipitations are more difficult to calculate both due to the heterogeneity of precipitations distribution over the sea as well as to the lack of data. The expected multiannual average input quantity of water from precipitations ranges between a minimum of 119 km^3 per year (Soliankin 1963; Bondar 1986) and a maximum of 300 km^3 per year (Unluata et al. 1990).

Table 5.3 Estimation of Black Sea waters balance elements in km²/year multiannual averages (inputs: Q from rivers, Pp from atmospheric precipitations, B through Bosphorus Strait, K through Kerch Strait; outputs: E through evaporation, B1 through Bosphorus Strait, K1 through Kerch Strait)

No.	Author	Q	Pp	B	K	E	B_1	K_1	Total inputs	Total outputs
1.	Leonov (1960)	309	230	193	95	365	392	70	462	462
2.	Berenbeim (1959)	340	120	193	59	280	398	34	432	432
3.	Rojdestvenski	294	254	229	38	301	485	29	514	514
4.	Şerpoianu (1973)	336	120	123	53	340	260	32	292	292

The estimation of inputs and outputs through Bosphorus Strait differs widely from one author to another.

The inputs through Bosphorus Strait in the Black Sea are assessed by Sverdrup, Neuman and Pierson (1966), Leonov (1960), Berenbeim (1959) at 192–193 km³/year, by Soliatkin (1963) at 176 km³/year, by Şerpoianu (1973) at 123 km³/year, by Bondar (1986) at 203 km³/year and by Unluata et al. (1990) at 312 km³/year, by Ozsoy and Unluata (1997) at 305 km³/year.

The outputs from the Black Sea through Bosphorus towards Mediterranean Sea are assessed by Spindler at 416 km³/year, by Sverdrup at 397 km³/year, by Neuman and Pierson (1966) at 462 km³/year, by Soliatkin (1963) at 340 km³/year, by Şerpoianu (1973) at 260 km³/year, by Bondar (1986) at 371 km³/year, by Unluata et al. (1990) at 612 km³/year, by Ozsoy and Unluata (1997) at 605 km³/year.

The inputs from Azov Sea are assessed between a minimum of 22 km³/year and a maximum of 95 km³/year (Leonov 1960), and the outputs from the Black Sea towards Azov Sea, between a minimum of 29 km³/year and a maximum of 70 km³/year (Leonov 1960).

The same large differences are also present in the calculation of outputs, by evaporation.

Below are described the estimates made by four authors (Table 5.3) on the understanding that they are only evaluation attempts of Black Sea waters balance, with the concern for a result to express a perfectly balanced balance.

Each of the balance components illustrates a monthly, seasonal and annual sharp variability, especially according to the climate conditions in the catchment area of the Black Sea and in the marine space.

We note also that in the evaluations carried by now, no importance is attached to the subsidence processes in the Black Sea basin which may favor additional water accumulations, modifying thus the value of the balance's negative components.

5.6 The Temperature of the Waters

The temperature of surface waters as well as the air temperature is characterized by significant differences between the two compartments, Western and Eastern, between which the central part with transition character is located.

The largest differences are recorded during winter (Fig. 5.1), those during summer being insignificant (Fig. 5.2). In the autumn, the cooling of surface waters starts from the North-West (from Fidonisi Bay) where in September is recorded an average temperature of 18.4 °C, towards the South-East corner, where the average temperature in September is of 21.2 °C (Table 5.4).

The surface waters from the Western compartment are characterized by the largest thermal heterogeneity due to the input of river waters and prevailing cold and dry winds, flowing from North-East.

The average temperature of surface waters is distributed according to model with thermal gradients proeminent towards the extremes of Western and Eastern basins and with small gradients in the central part.

The waters temperature in vertical profile presents a stratification specific for the Black Sea.

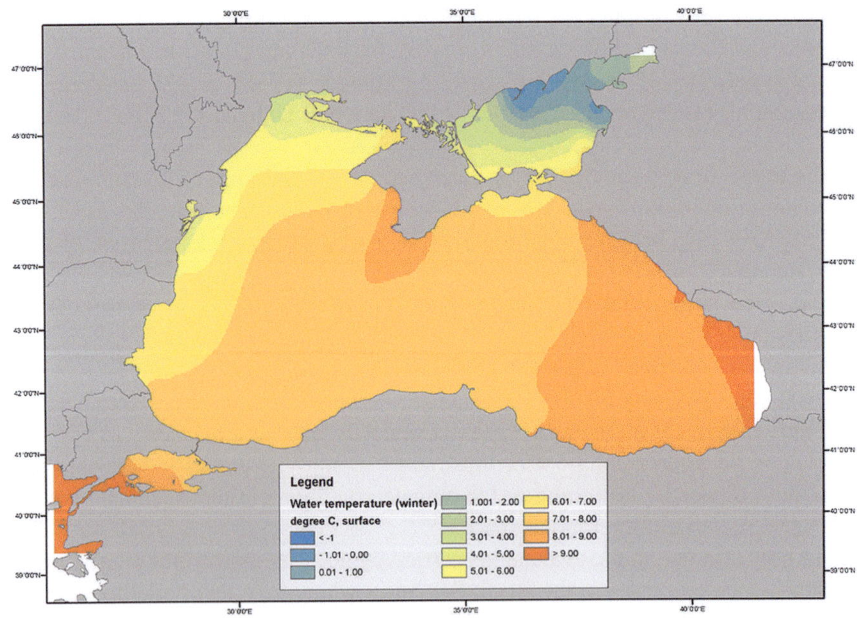

Fig. 5.1 Annual average temperature of Black Sea surface waters during winter (after World Ocean Atlas)

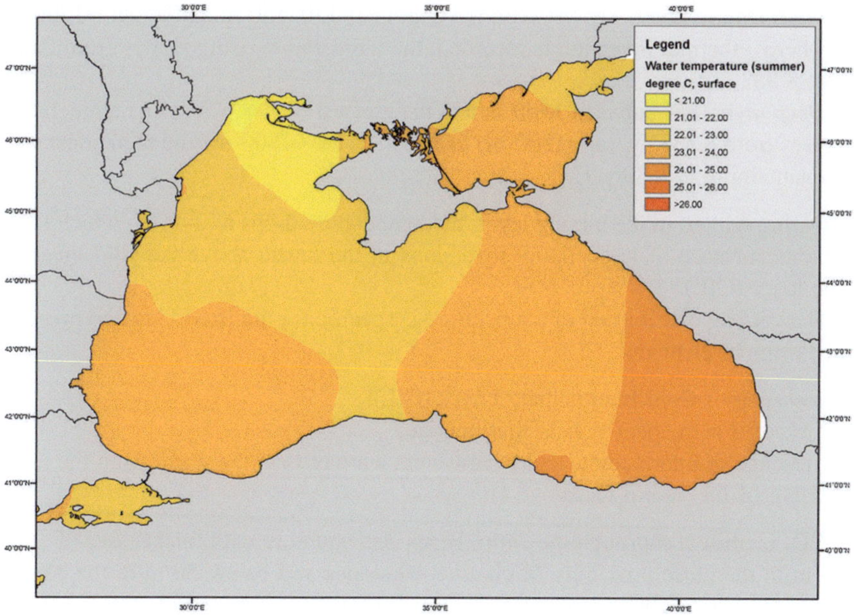

Fig. 5.2 Annual average temperature of Black Sea surface waters during summer (after World Ocean Atlas)

Table 5.4 Distribution of Black Sea surface waters temperature during winter and summer (according to Black Sea GIS data)

Month	Sector	Water temperature in Western compartment (°C)	Water temperature in central compartment (°C)	Water temperature in Eastern compartment (°C)
I	N	Below 2–7	7–8	7–10
I	S	7–8	6–7	8–12
VIII	N	21–22.5	22.6–22.9	22.6–22.8
VIII	S	22.5–23	22.6–22.9	22.6–22.8

In the deep basin of the Black Sea, during summer the following average thermal stratification is registered:

- *mixing layer*, between surface and 10–12 m depth, with differences between surface and lower limit of 1–1.5 °C;
- *seasonal thermocline layer* between the 10–12 and 40–45 m depths, where the thermal gradient reaches 12–14 °C;
- *Cold Intermediate Layer (CIL)* lies between the depths of 40–45 m and 130–150 m, where the temperature drops by 1–1.5 °C;

- *Intermediate layer*, lies between the depths of 130–150 m and about −1500 m, where a thermal inversion is recorded, the temperature rising slowly from 8.000 to 8.883 °C;
- *Deep layer*, lies between 1500 m and the greatest depths, where the temperature rises from 8.883 °C (at −1500 m) to 8.896 °C (at −1800 m) and then constantly maintaining at 8.896 °C.

During winter, in the mixing layer, the temperature drops to 2–8 °C, which leads to the occurrence of temperature inversions in the layers above the cold intermediate layer (CIL).

The vertical distribution of temperature is specific for the Black Sea and presents three original elements:

- presence of Cold Intermediate Layer (CIL);
- lack of a permanent typical thermocline;
- presence of intermediate and deep waters warmer than the water from the same depth of the World Ocean.

CIL is present thorough the entire Black Sea aquatory with thicknesses of 100–120 m in the close proximity of continental slopes and below 50 m in the central part of the basin. The maximum thickness is of 140–150 m and is recorded off to the continental slopes from the South-Eastern corner of the Black Sea. The CIL thermocline interface lies mainly at 25–30 m, except for the Southern and South-Eastern part where it drops at 40–45 m.

The intermediate and deep water layers have higher temperature than normal temperatures at the same depth, in relation with the input of warmer Mediterranean seas, through Bosphorus, as well as due to the geothermal caloric transfer at sea bottom level.

5.7 The Salinity of the Waters

The average annual salinity at surface ranges between a minimum of 14.5%, off to the Danube Delta, and a maximum of 20.7% in the central part of the basin (according to *SeaDataNet JRA6*). In front of Danube's mouths, the salinity drops to 5–8% due to the large volume of river water. The average annual salinity at surface has a longitudinal distribution, with low values and high gradients on the Western and Eastern sides and with values exceeding 18.20% in the central part of the basin where the **seasonal variations of the salinity at water surface** record high values.

- during winter, the minimum values are of 15.44% and are recorded off to the Danube Delta and in Odessa Bay; most part of the sea surface from the Western compartment has a salinity of 18.19–18.28%, the Central-Eastern part has waters with maximum salinity of 18.28%, and the Caucasus waters have 17.97% (Fig. 5.3);

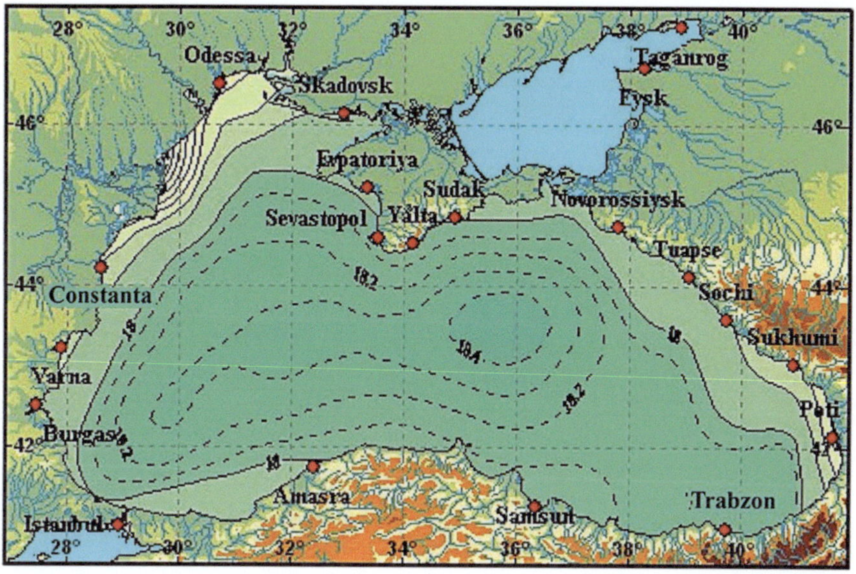

Fig. 5.3 Distribution of Black Sea surface waters salinity during winter. *Source* SeaDataNet JRA6

- during summer, the minimum salinity is recorded in front of Danube Delta and Odessa Bay, with values ranging between 14 and 15%; the central part of the basin, the salinity ranges between 18 and 18.50% and on the Caucasus coast between 16.9 and 17.5% (Fig. 5.4);

The seasonal variations are clearly related to the input of river waters towards Black Sea and the ratio between atmospheric precipitations and evaporation.

The variation of vertical salinity in the deep basins follows a trajectory specific for the Black Sea basin, different from the rest of the World Ocean by the wide expansion of halocline.

The vertical profile of salinity distribution in deep basin presents three different floors (Stanev and Beckers 1999; Bakan and Büyükgüngör 2000; Oguz et al. 2001):

- the mixing layer, between 0 and −30 m, with salinities ranging between 18.0 and 18.25%;
- the halocline layers, which lies between −30 m and −200 where the salinity rises from 18.25 at 21.5%, with a gradient of 3.25%;
- Intermediate layer, lies between −200 m and −1000 m, where the salinity is rising slowly reaching 22.05% at −500 m, 22.20 at −700 and 22.29% at −1000 m.
- Deep layer, lies below −1000 m where the salinity is rising very slowly, reaching 22.32 at −1500 m, value which is maintained constant up to the greatest depths.

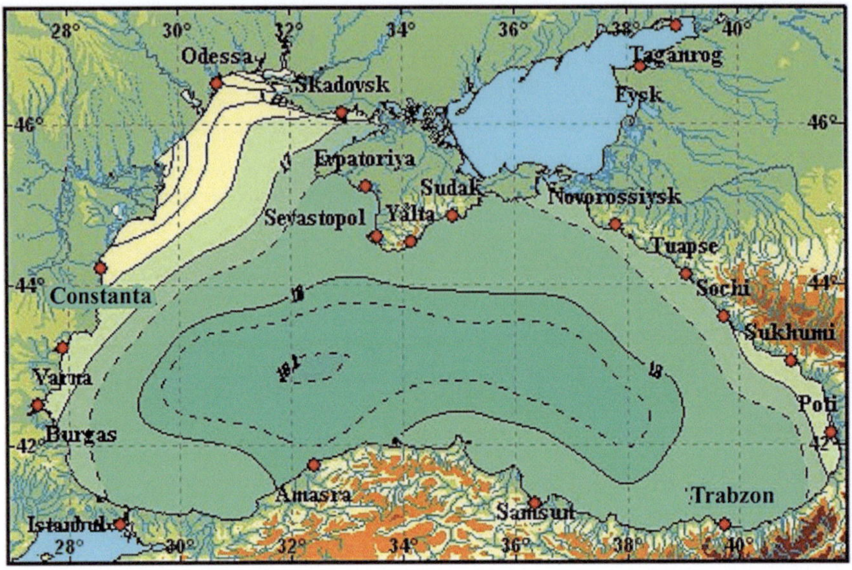

Fig. 5.4 Distribution of Black Sea surface waters salinity during summer. *Source* SeaDataNet JRA6

The vertical distribution of the salinity in the deep basin is the result of multiple diffusion processes and mixture of water masses with different origins, as we will show further.

5.8 The Density of the Waters

The sigma-t density presents values rising with depth, determining, to a large extent, the vertical distribution of main chemical components, with all the required processes.

The vertical distribution of density presents the following levels (Oguz et al. 2001; Karakas et al. 2002):

- Superficial layer which lies between the surface and −10 up to −15 m, with sigma-t densities of 10.5–11 kg/m^3;
- Permanent pycnocline layer which lies between −15 m and −120 m where the sigma-t density rises from 11–11.5 to 15.5–16 kg/m^3, with a gradient of 5–6 kg/m^3;
- Deep layer which lies at depths greater than 120 m, with sigma-t density of 16–16.5 kg/m^3.

We mention the importance of firm pycnocline presence with high gradient which has a decisive influence on the development of vertical transfer processes (Gregoire and Stanev 2001; Konovalov and Murray 2001). However, the intense wind activity during winter generates frequently convection processes which determine the local ventilation of pycnocline and of deep waters up to depth of 400–500 m (Gregoire and Stanev 2001).

5.9 Oxygen Layers

The dissolved oxygen register maximum values in the superficial layer between surface and −10 m where it ranges between 330 and 380 μM/l (micromoles per litre) (*Black Sea GIS*, 1998). At 100 m depth, the oxygen quantity drops until below 5 μM/l in the central part of the deep basin, maintaining however at values of 275–300 μM/l in the close proximity of continental slopes. Below 150 m, the oxygen drops fast and disappears between −190 and −200 m. The vertical distribution of oxygen quantity from the Black Sea deep basin follows the model below (Aubrey et al. 1992; Konovalov and Murray 2001; Oguz et al. 2001):

- The superficial layer, which lies between 0 and −5 m, where the most intense exchanges with the atmosphere take place, is characterized by large quantities of dissolved oxygen ranging between 350 and 379 μM/l.
- The mixing layer, which lies between −5 m and 30–35 m depth, where the oxygen is at 90–110% saturation ranging between 305 and 350 μM/l. Here the oxygen comes from the atmosphere, from the gas exchange and the photosynthetic activity of phytoplankton ensuring the primary production. The oxygen quantity from the mixing layer ranges seasonally according the temperature, salinity and the degree of surface roughness which determine the oxygen solubility and the development of primary production processes.
- The oxycline layer, which lies between 35 and 60 m depth, where the quantity of dissolved oxygen drops from 300 to 5–6 μM/l. The decrease in oxygen quantity is due to its active consumption in the oxidation processes of organic matter and sulphides (Konovalov and Murray 2001).
- Oxic/anoxic layer, called also suboxic layer (SOL), located within the permanent pycnocline, between 55–60 m and 100–120 m, is close relation with pycnocline, where the quantity of dissolved oxygen drops from 5 to 6 μM/l to values close to zero. In the deep basin, the oxygen disappears at the depth of 190–200 m. In the suboxic layer (SOL) a water mass is formed specific for the Black Sea, as we will see further.

The vertical distribution of dissolved oxygen quantity ranges from one place to another according to the dynamics of water masses so that we cannot identify consistent repartitions for the entire Black Sea basin. We may speak about a relief, with hills and depressions, of the upper surface of the anoxic layer, as we will see further.

5.10 Hydrogen Sulphide and Sulphides

Andrusov (1890) discovers the presence of the water layer rich in hydrogen sulphide and free of oxygen, during the expedition on board of Cernomoreţ vessel, deepening afterwards its investigations in more expeditions (Andrusov 1918, 1926). Later, Arhanghelski (1927) and Arhanghelski and Strahov (1938) mention the general chemical structure of the Black Sea where the waters with H_2S have a prevailing role (Fig. 5.5).

After the World War II detailed researches appear on the reduction process of sulphates (Sorokin 1982), the production process of H_2S (Deuser 1971), the dome-shaped anoxic-oxic interface (Skopintsev 1975). Later the suboxic layer (Murray et al. 1989), the vertical oscillation of anoxic interface (Konovalov 1995; Konovalov and Murray 2001), the redox cycles developed at the level of anoxic-suboxic interface (Oguz et al. 2001), the sulphur budget of the Black Sea aquatory are discovered (Neretin et al. 2001).

The vertical distribution of the average quantity of H_2S (Neretin et al. 2001) illustrates a continuous rise from 150 m depth up to −2000 m (Table 5.5).

The analysis of production distribution and H_2S consumption showed the existence of an horizon with maximum production between 500 and 1000 m depth, of a maximum consumption horizon higher than −500 m and a neutral horizon below −1000 m (Neretin et al. 2001).

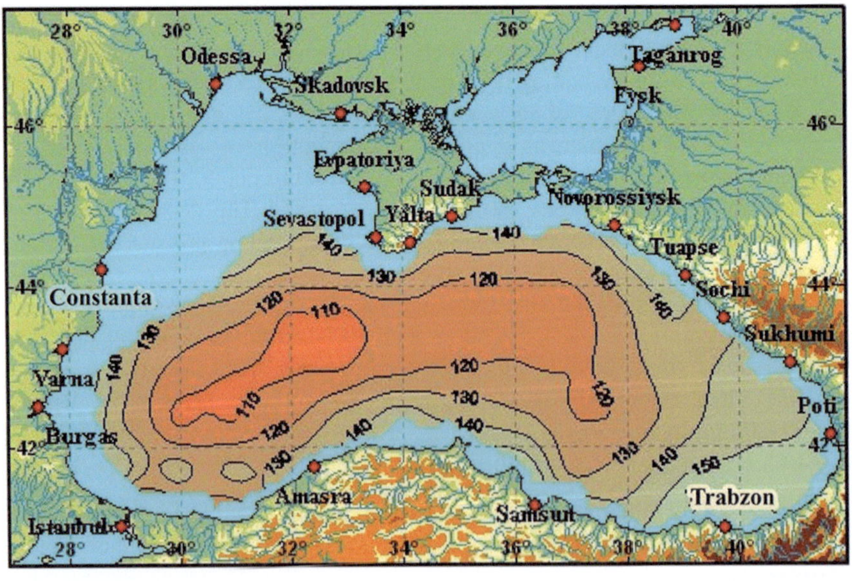

Fig. 5.5 Distribution of H_2S in Black Sea basin. *Source* SeaDataNet JRA6

Table 5.5 Quantity of hydrogen sulphide in the Black Sea (after Neretin et al. 2001)

Layer (m)	Average quantity of H_2S (μM/l)	ÓH_2S 10^9 (t)	ÓH_2S (mol/m^2)
100–200	21	0.022	2.0
200–300	66	0.068	6.5
300–500	140	0.283	28.1
500–1000	274	1.315	132.1
1000–1500	339	1.469	159.1
1500–2000	368	1.220	150.7
2000–2200	368	0.210	40.7
Total		4.587	519.2

The H_2S formation starts with the advection of suphur to water surface which ranges between 0.8 and 200 Tg/year according to the intensity of turbulence and advection processes (Neretin et al. 2001).

In the suboxic layer (SOL) sulphur oxidation processes take place in two phases: at the beginning, the sulphurs are oxidized to elementary sulphurs and thiosulphates; then, the thiosulphates are oxidized to sulphates with the help of Chemolithoautotrophic bacteria, especially those of *Thiomicrospira* sp. type. The sulphate formed as above is reduced by the formation of hydrogen sulphide. The role of biogeochemical process in the formation of H_2S has the same importance as the role of strictly abiotic processes, fact proved by the increase of Bacteriochlorophyll e at oxic-anoxic interface.

After the formation of H_2S at the chemocline level, it may accumulate or may be decomposed by oxidation:

$$H_2S + 2O_2 = SO_4 + 2H$$

The presence of sulphur in water mass and sediments is due to the formation processes of pyrite type colloidal and metha-colloidal iron biosulfide (FeS_2) by the decomposition of organic remains under anaerobic conditions (Cutter and Klukholm 1999; Konovalov and Murray 2001; Oguy et al. 2001; Neretin et al. 2001). Under aerobic conditions, the iron sulfides are instable, easily oxidizing in ferrous sulfate which in the presence of the oxygen passes in ferric oxide. This, in its turn, decomposes quickly by hydrolysis, resulting the iron hydroxide and free H_2SO_4 which passes in solution. Therefore, 60% of the suphides are formed in the upper floor of the anoxic layer (Lein and Ivanov 1991). In sediments significant quantities of colloidal iron biosulfide accumulate which shall ensure a continuous hydrogen sulphide towards the aquatory.

The model of development of H_2S production and consumption processes (Fig. 5.6) illustrates the presence of three levels:

- the H_2S production level by the reduction of sulphates in the presence of Chemolithoautotrophic bacteria, located in the anoxic layer and sediments;
- the H_2S consumption level in suboxic layer (SOL) by sulphides oxidation;

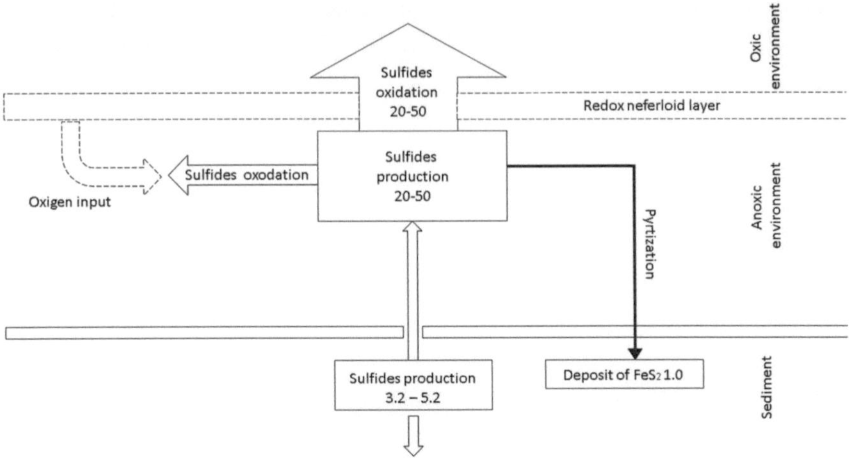

Fig. 5.6 Model of the development of H_2S production and consumption process in Black Sea aquatory. The arrows thickness illustrates the process magnitude on that line. The H_2S quantities are expressed in tonnes/year (after Neretin et al. 2001)

- the oxic level which is the source of organic matter particles and organic carbon resulted from the primary production, which shall provide the basis of biosulphides formation processes. At the same time, large quantities of H_2S shall be consumed here through oxidation.

In the budget of hydrogen sulphide of Black Sea an important role is played by the intrusion of Mediterranean water plume through Bosphorus Strait.

The Mediterranean origin waters come out on the continental shelf in front of Bosphorus Strait, load with oxygen, submerge on the continental slopes, entering through multiple diffusion in the water mass of anoxic layer. Here they determine oxidation processes of sulphides, with the reduction of H_2S.

Therefore, the Black Sea aquatory is a chemical system with self-adjustment, where the H_2S input processes are controlled through feedback by the H_2S consumption processes (Fig. 7.6). This explains the stability of oxic/suboxic/anoxic interfaces level, at least during the last 100 years. The dynamics of anoxic-suboxic relief interface, with the lateral movement of some hills in the place of depressions, leads to the wrong conclusion according to which the anoxic layer is getting close to the surface, fact which is not scientifically proven.

5.11 Water Masses

The water masses were identified as large water volumes with certain specific thermal, saline and density properties, formed in some genetic centers, which may travel long distances without changing their properties. These water masses maintain their features for very long periods. The concept applies to oceanic waters where differentiation conditions of such water volumes exist.

In the Black Sea we may speak, somewhat inappropriately, about water masses identified after their physical and chemical properties.

The water masses of the Black Sea are known due to the researches carried out in the last 20 years and, in particular, to the HYDROBLACK 91 programme and Cooperative Marine Science Programme for the Black Sea, as well as to recent researches (Belyaev et al. 1997; Cutter and Klukholm 1999; Huang et al. 2000; Oguz et al. 2001; Neretin et al. 2001).

In present, there are several relative options on the presence of water masses in the deep Black Sea waters.

Oguz et al. (1994) describe four water masses:

(1) surface waters, which lies in the layer located between 0 and −40 m strongly affected by the exchanges with the atmosphere, where two layers were subdivided: a superficial layers which lies between 0 and −25 m with temperatures of 3–22 °C according to the season, with salinities of 17–18% and dissolved oxygen close to saturation; a lower layer which lies between −25 and −40 m where the seasonal thermocline is recorded with a thermal gradient up to 14 °C;
(2) the waters of Cold Intermediate Layer (CIL) bordered by the hydro-isothermes of 8 °C, with temperatures of 7.10–7.20 °C, salinities of 17.5–18.2% and an increasingly smaller quantity of dissolved oxygen; these waters are located between −40 m and −150 m, with some depth variations according to the basin's sector to which we refer; in this layer there is the main halocline and pycnocline, as well as the suboxic layer (SOL);
(3) intermediate waters where the temperature rise from 8.50 °C to 8.80–8.82 °C, salinity is maintained at values of 22.100–22.125%, oxygen is missing and H_2S remains the prevailing gas dissolved in water; these waters are quartered between −150 m and −1800 m;
(4) bottom homogeneous waters (*Bottom Homogeneous Layer*), below −1800 m depth, with constant temperature and salinity of 8.820 °C, namely 22.125%.

These masses are not perfectly stable but they have a certain dynamics in time and space. The entire Black Sea aquatory presents the dynamics specific to water masses from medium and small semi-enclosed basins (S.G. Boguslavski si colab., 1992), so that the old theory which treated the Black Sea aquatory as unaffected by vertical exchanges is no longer valid.

Gregoire and Stanev (2001) consider that there are four water masses identified in the basis of biogeochemical criteria:

(1) Surface waters;
(2) Oxicline and upper nitrocline layer waters;
(3) Transition waters corresponding to suboxic layer;
(4) Deep anoxic waters

The authors consider that we may speak about only two main water masses:

- Surface waters, located between 0 and −200, with a total water volume of about 30,000 km^3, average salinity of 18% and density between 10.5 and 16 kg/m^3;
- Deep water, which lies below −200 m, with a total volume of about 500,000 km^3, average salinity of 22% and density of 16–17.5 kg/m^3;

Between these two water masses there are bilateral exchanges, meaning that about 700 km^3 of water pass annually from the deep waters to surface waters and about 524 km^3 of water pass annually from surface waters to deep waters. Irrespective of whether we accept the presence of several water masses or of only two water masses mentioned above, in the Black Sea are present three water layers with original properties specific only to this sea. These are the cold intermediate layer (CIL), suboxic layer (SOL) and anoxic layer. To these are added the Mediterranean water masses which enter through Bosphorus Strait in the Black Sea.

The Cold Intermediate Layer (**CIL**) was identified by Spindler (1890) and analyzed in details by Spindler and Wrangel (1895). Many years later, Antipa describes it as "minimum temperature layer", mentioning its important role in the structure and operation of pelagic biocoenosis. CIL is known also as Cold Intermediate Waters or CIW. CIL is characterized by a thermal inversion, being bordered by the hydroisotherm of 8 °C. CIL encompasses waters with minimum temperature of 7.10–7.20 °C and salinity of 17.5–18.2%. The genesis of CIL was explained differently in the course of time.

During the interwar period, Arhanghelski and Strahov (1938) consider that CIL is formed during winter, when the cold surface waters submerge. The convection processes are favored by very cold and windy winters from the North-Western compartment of the Black Sea. Sverdrup et al. consider that the massive input of freshwaters from rivers and atmospheric precipitations determines all the properties from the upper layer of the Black Sea, including the formation of CIL.

More recently, Filipov (1965), Tomalzin (1985) consider that CIL is formed in two main genetic centers: the North-Western shelf and the shelf in front of Kerch Strait, the input of river waters being essential. During the same period, Ovchinnikov and Popov (1987) claim that CIL is formed in the center of cyclonic vortexes generated by currents through local convection. Ozsoy and Unluata (1997) consider that the advection and convection processes developed in the North-Western basin of the Black Sea are insufficient to form the CIL. Ivanov et al. (1997) support the assumption of I.M. Ovcinnikov and Y.I. Popov on CIL formation in the cyclonic vortexes from the deep basin during winter, considering less important the role of advection process from the North-Western basin.

Karakas et al. (2002) reexamine the issues of genesis based on the last oceanographic researches jumping to the conclusion that the waters of North-Western shelf represent the genetic center of CIL from where they reach the deep basin through advection. Here, the CIL waters are circulated by secondary anticyclonic currents from the periphery of Black Sea Rim Current, reaching in its vortexes, as well as in the remaining basin.

The Suboxic Layer (or SOL), was identified and studied recently (Murray et al. 1989).

SOL is individualized as a very stable layer between the lower part of the oxycline, with an upper limit at 55–60 m depth and a lower limit at −120 m, rarely to −180 m. SOL is the layer where the oxygen quantity drops, without disappearing and the H_2S quantity rises. The structure and maintenance of SOL is provided by three mechanisms (Oguz et al. 2001):

- Sulphides oxidation and nitrogen transformation in close relations with the iron and magnesium cycles;
- Anoxygenated photosynthesis processes through the activity of anoxygenic phototrophic bacteria which lead to the phototrophic reduction of HS, Mn and Fe, as well as the carbon dioxide. This mechanism was proved by ascertaining the decomposition of large quantity of bacteriochlorophyll pigments. The pigments come from bacteria which actively develops under extremely reduced illumination condition;
- the H_2S oxidation through SOL penetration of water plumes richer in oxygen.

The anoxic layer lies below −200 m and has a water volume of about 500,000 km^3, until recently considered a stable and inert water mass, with no dynamic activity. Recent researches (Karakas et al. 2002) proved the existence of a clear dynamic stratification in the Black Sea deep basin which affects the anoxic layer:

- between surface and −100 m the Black Sea Rim Current flows, with cyclonal direction, which is felt up to −500 m, as shown below;
- between −500 m and −1800 m a slow anticyclonic current flows;
- below −1800 m a very slow cyclonic current exists.

The entire dynamic activity from the anoxic layer is due to the geostrophic processes, density discontinuities, diffusion of Mediterranean seas, as well as to the convective movements due to the geothermal flow felt in the water layer located at 300–500 m above Black Sea bottom.

The Mediterranean water masses enter in the Black Sea through Bosphorus Strait, as shown above, in the sub-chapter designed to the analysis of this strait. At the exit from Bosphorus Strait, the Mediterranean waters enter in Black Sea with a volume of about 300 km^3/year, meaning a flow of roughly 10,000 m^3/s. These waters have a salinity of 35.50% and temperature of 14.5 °C (Ozsoy and Unluata 1997).

The Mediterranean waters descend on the South-Western shelf, are cooling from 14.5 to 8 °C above the edge of continental shelf and their salinity drops from 35.5 to 22.8%. As these waters descend along the continental shelf, intrusions are formed in the depths which are pushed in the water mass towards the center of the deep basin, recorded for the first time by Bogdanova (1969), Boguslavsky et al. (1992).

Recent researches elucidate the intrusion and diffusion processes of Mediterranean seas in the South of Black Sea (Oguz et al. 1994; Ozsoy and Unluata 1997). The unitary model illustrates a unique process which starts with the mixing of Mediterranean waters above the shelf in front of the Northern exit of Bosphorus Strait, followed by a double diffusive convection. As the Mediterranean waters flow along the continental slope, the warmer and saltier water intrusions develop horizontally (Fig. 5.7). Mediterranean water lenses appear frequently in the Black Sea deep waters, which are largely spread intrusions.

The mixing of water masses and ventilation in the Black Sea deep waters are recently known processes. Previous approaches, essentially fixist, considered the entire anoxic water mass as unaffected by vertical exchanges due to the very strong pycnocline which separates water layers with different densities. The researches carried out on vertical transport through halocline and pycnocline proved the existence of efficient and quite fast vertical ventilation. Therefore, Buesseler and Livingston (1972), who analyzed the diffusion of radioactive pollutants generated by the nuclear disaster of Cernobal, noted a fast vertical transfer between the surface and about 500 m depth. To this are added the diffusion processes from the turbulent layer along the continental slopes, as well as the convective processes from the deep layer, located below −1800 m due to the geothermal flow (Fig. 5.8).

Briefly, in the present day stage of knowledge, it is considered that the Black Sea waters are affected by ventilation and mixture processes developed throughout the aquatory with different intensities, directions and rhythms:

- the superficial waters of Ekman layer are affected by intensive ventilation processes due to the strong interactions between atmosphere and sea;
- the pycnocline and halocline waters are affected by vertical mixing processes which are felt up to −500 m depth;

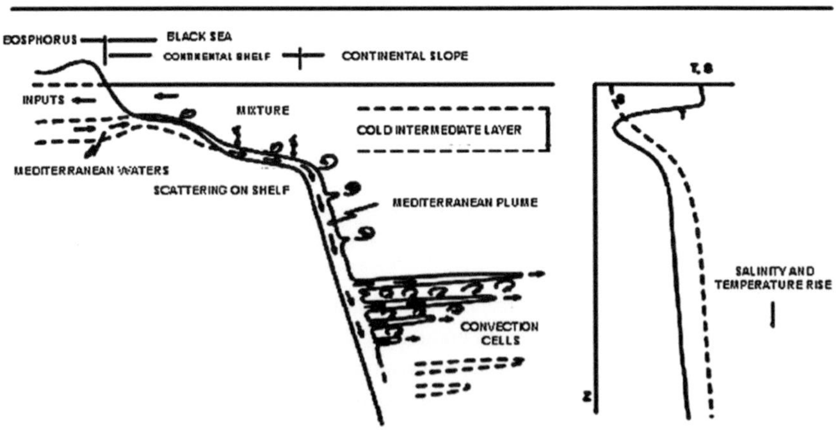

Fig. 5.7 The mixing processes generated by the entering of Mediterranean seas in Black Sea through Bosphorus Strait (after Ozsoy and Unluata 1997)

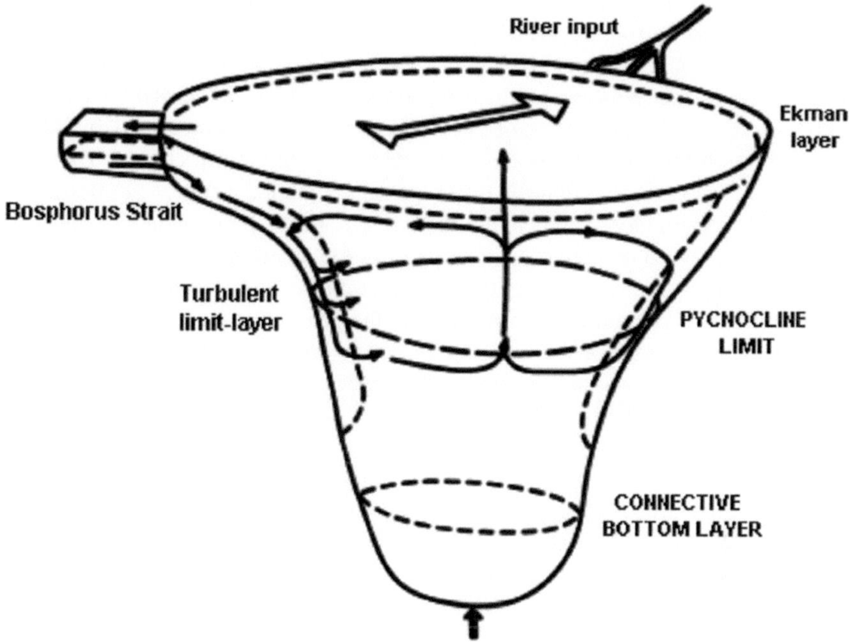

Fig. 5.8 Model of recirculation and mixing paths of Black Sea waters (after Ozsoy et al. 1993, 1997)

- waters between −500 and 1800 m are closely connected to the processes developed along the continental slopes in the turbulent limit-layer;
- waters below −1800 m up to the greatest depths (convective bottom layer) are affected by convection processes generated by heat geothermal flow.

5.12 The Currents from the Black Sea

The currents from the Black Sea were subject to oceanographic researches even since the second half of 19th century, when Ch. Hartley, N.I. Andrusov, I. Voeikov, Wranghel, G. Antipa perform the first observations and measurements. In the first half of 20th century, N.M. Knipovici, O.I. Schmidt, P. Fundăţeanu, R. Ciocârdel and G. Antipa, continue the researches preparing a model improved later by Neumann. Antipa (1941) is the first Romanian author who draws up a synthesis map of surface circulation from Black Sea based on the data existing at that moment, starting with the measurements of Ch. Hartley, of the Russian authors (especially N.M. Knipovici) and of the Romanian authors R. Ciocârdel, N. Gavrilescu, P. Fundăţeanu).

According to G. Antipa synthesis, in the Black Sea a major cyclonal circulation develops which surrounds the deep basin concentrating above the isobath of −180 m. From these current two cyclonal vortexes are detached in the deep basin, a smaller one in the Western basin and a broader one in the Eastern basin. The circulation from coastal waters of North-Western shelf is represented by Antipa (1941) through three permanent currents detached from the rim current. The study of Antipa (1941) and his synthesis map represented the main source of information for the Romanian readers influencing the entire relevant geographic thinking until today. During the same period Neumann draws the attention on the complexity of Black Sea current issues, highlighting the importance of geostrophic phenomena. The analysis of dynamic relief which led to the production of an absolute topography map of sea surface allowed the author to produce a more precise current map. Later, three approaches of Black Sea currents appeared in Romania which make the confusions permanent.

The first one is owed to Pora and Oros, who take over the map of Gr. Antipa, supplementing it by a "*permanent*" current in the North-West of the basin.

At the same time, Ujvary publishes (without mentioning the source) a map where two overlapped currents are presented in the deep basin: superficial current, which flow at surface, and submerged current, with opposite direction, considered compensation currents, in depth. The third recent source of confusion we find in the *Geography of Romania*, Vol. I, Physical Geography (Romanian Academy Publishing House, 1983, p. 369) where M. Schram and M. Pantazică make short work of the complex issue of surface circulation in Black sea in two phrases which ignore totally the relevant oceanographic achievement. In the same source, we find the same confusing approaches related to the coastal circulation (Selariu, p. 370).

These approaches of permanent circulation in the deep basin, which have no theoretical and at least of all a practical basis (are not based on measurements data) were taken over, unfortunately, in some manuals of Romanian geography (XII grade), generating serious confusions. The same confusions are generated also by the presentation of coastal circulation, where the permanent currents, which in fact do not exist, were given names: Crimean Current, Odessa Current, Dobrogea Current.

All these imposed a certain way of thinking accepted by the geographists and disseminated by manuals, courses or other mass media means even nowadays. This "classic" way of thinking requires the acceptance of two categories of permanent currents:

- circular currents from deep basin, which were presented closely to reality by Gr. Antipa;
- coastal water current from the Western basin.

The latter were considered as permanent on general North-East-South-West paths, having an essential role in the scattering of Danubian alluvia towards South with the formation of the marine part on Danube Delta, of lagoon barriers as well as the shores and beaches in the South of Midia Cape. The idea of the existence of permanent currents in the North-West of the Black Sea was erroneously taken over from Sir Hartley (1862), Chief Engineer of the Commission of the Danube River and largely publicized by Antipa (1910, 1937, 1941). The reality is that Sir Ch. Hartley had made a clear presentation on the oscillating character of the circulation from North-West of the Black Sea: "On the Danube Delta coast there is a coastal current which, during regular winds from North and North-East and during periods of calm, flows constantly towards Bosphorus, so towards South. During the strong winds from South and South-West this current disappears and is replaced by a very slow surface current from South to North" (Hartley 1862).

The researches made after 1970 as well as the international efforts within CoMSBlack and EROS programs developed after 1990, proved the inexistence of permanent currents oriented North-East to South-West in the coastal waters from the Western basin and defined the structure and real operation of surface circulation in the Black Sea (Oguz et al. 1994; Ozsoy and Unluata 1997; Oguz and Besiktepe 1999; Ginzburg et al. 2002 and various other papers).

The actual stage of knowledge, based on tens of thousands measurements, on analysis by remote detection and GIS techniques, it is accepted the existence of a very original circulation in the Black Sea due to the convergence of several factors:

- the horizontal and vertical differences in density, temperature and salinity
- differentiate exposure towards the prevailing winds due the presence of asymmetrically spread orographic barriers;
- development of some geostrophic type processes, eddy type, as well as the Rossby baroclinic processes;
- prominent morphological differences of submarine relief from various parts of the basin, especially the presence of continental shelf widely spread in the North-West of the basin;
- the exchange of waters with Mediterranean Sea and Azov Sea.

Recent researches proved the existence of several types of circulation in the Black Sea basin: major surface circulation in the deep basin, sub-basinal surface circulation and mesoscale, upwelling and downwelling vertical circulation.

Major surface circulation in the deep basin. The model accepted now, which we may call the "Oguz model" (Fig. 5.9) illustrates the presence in the deep basin of a cyclonal current called Black Sea Rim Current. This current flows actively in the layer between the surface and −150 m, has a geostrophic character and speeds of 15–20 cm/s at surface and over 20 cm/s in the core. The current speed drops as the depth is greater, being felt up to −500 m, where a slow circulation exists, with maximum speeds of 2 cm/s, very variable, with countercurrents and vortexes (eddy) to which the recycling cells are related (Kukushkin 1992; Oguz et al. 1994; Oguz and Besiktepe 1999; Bakan and Büyükgüngör 2000). The total average water transport is

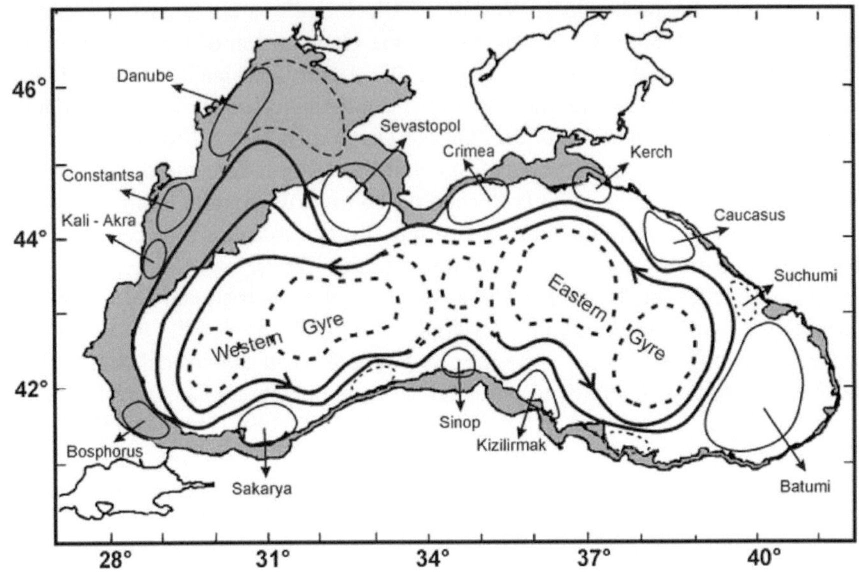

Fig. 5.9 Circulation pattern of the surface layer of the Black Sea (after Oguz et al. 1993; *Korotaev*, Oguz et al. 2003)

of 5.75 million m^3/s (5.75 Sv), with important seasonal variations (Stanev et al. 2002). During winter, the total transport of water is of 6 Sv, in spring of 8 Sv, summer—4 Sv, and autumn—5 Sv (Sv: Sver-drup; 1 Sv = 1 million m^3/s).

The Rim Current has 50 km width, doubles slightly with wave width of 150–200 km and flows into the deep basin, beyond the shelf edge, above the continental slopes. In transversal section horizontal shearing processes appear frequently, which determines the formation, towards the exterior of 9 well shaped anticyclonic vortexes called Sevastopol, Kaliakra, Bosphorus, Sakarya, Sinop, Kazalirmak, Batumi, Caucasus and Crimeea, to which are added several small, secondary vortexes (Fig. 7.9). The Rim Current separates the coastal waters with surface salinity of 15.44–17.97% from the deep basin waters with surface salinity of 18.54–19.00%.

The temperature of surface current waters is higher than that of coastal waters with 2–3 °C. Under the level of seasonal thermocline, the temperature drops due to the presence of waters from the cold intermediate layer (CIL). Within the big ring of Rim Current two cyclonic vortexes are formed, Western and Eastern, with several internal vortexes with a very active dynamics, continuously changing in time and space.

The entire spatial and temporal structure of the Rim Current is determined by the geostrophic processes, by doubling and eddy, being discontinued in time and very variable in space.

The genesis of Black Sea Rim Current was initially connected to the general circulation, with cyclonic character, of the atmosphere above the Black Sea. Recent researches show the existence of a seasonal thermohaline circulation which affects the entire water layer between surface and 500 m depth. This thermohaline circulation is accelerated by winds on cyclone trajectories. The buoyancy anomalies generated by the river input, atmospheric precipitations and evaporation contribute to the development of cyclone model of Black Sea Rim Current.

The sub-basin circulation develops in coastal regions and is more active in the Western compartment, where the most extended continental shelf exists, as shown above. The researches carried out in Romania by G. Şerpoianu, C. Bondar, I. Nae, V. Diaconu, M. Rădulescu, I. Postolache, V. Malciu, between 1970 and 1985, and E. Vespremeanu, during 1980–1990, proved the absolute importance of three factors controlling the surface flow in the waters of North-West shelf of the Black Sea: wind direction and intensity, mixing of Danubian river waters with the marine water, coastal relief.

Recently, the results of the researches made in the last 15 years appear and specify the structure and operation of coastal water circulation, especially of those from North-West of the Black Sea (Simonov and Altman 1991; Mikhailova and Shapiro 1996; Sur and Ilyin 1997; Oguz and Besiktepe 1999; Stanev and Beckers 1999; Stanev and Staneva 2000; Staneva et al. 2001; Bekers et al. 2002; Karakas et al. 2002; Stanev et al. 2002).

The coastal marine aquatory lies between the water line and a limit situated above the shelf edge which, under the conditions of North-West of the Black Sea, is located at about −150 m. Between these limits, four categories of waters are identified based on the hydrodynamic behavior: shelf waters, littoral waters, shore waters and beach waters.

The shelf waters are overlapped to the continental shelf, between a lower limit located at about −150 m and an upper limit at −40/−45 m. In these waters prevails the circulation determined by wind with the development of Ekman stratification as well as the geostrophic processes. The specific hydrodynamic is reflected in the horizontal and vertical dynamics of alternative wind currents (Şerpoianu 1981). The exterior boundary (to the high sea) of the compartment is influenced by the Black Sea Rim Current from which several small, anticyclonic *eddy* cells detach which enter discontinuously in the shelf waters, even though the North-Western shelf waters are decoupled from the main circulation in the deep basin (Oguz et al. 1994; Oguz and Besiktepe 1999). In the shelf waters the cyclone or anticyclone *eddy* cell occur frequently due in particular to the wind activity (Oguz and Besiktepe 1999).

The littoral waters are overlapped to the slope and littoral glacis, between depths of −40/−45 m and −10 m. This type of waters are affected by the specific processes of frictional flow generated by wind, presenting a prominent variability in time and space (Şerpoianu 1973; Bondar et al. 1973). The Ekman Stratification is maintained as resulting from the excellent analysis presented in the *Atlas of Oceanography*. The winds from North and North-East direction determine a North-South oriented general circulation ranging between 0.3 cm/s and 3.6 cm/s, which led to the movement of a water flow ranging between 5750 and 77,559 m^3/s. This circulation with South direction alternates with the North circulation determined by winds flowing from the Southern sector and forming currents with speeds of 0.1–2.9 cm/s, activating a water flow of 2687–61,586 m^3/s. *Therefore, there is an alternative circulation from the Northern and Southern sector, with moderate prevalence of the Northern sector. In vertical profile, the currents vectors register decreases in speeds and shifting of directions according to the laws of Ekman.*

The shore waters lies between a lower limit overlapped to the shore terrace edge, at the depth of −10 m and an upper limit at the level of the proximal ditch from the shore profile located at about −1.5 m. The hydrodynamic regime of waters is dominated by the wave breaking processes and waves transformation by refraction, diffraction and reflection (Vespremeanu 1987). The waves refraction on the bars crests located on the shore terrace determines the formation of longshore currents which flows through the ditches between the bars. These currents developed in the coastal waters have alternative directions, from the Northern sector and the Southern sector, according to the orientation with which the waves train approached the shore. Frequently, rip currents are formed with an important role in modeling the shore relief (Vespremeanu 1987).

The beach waters lies above the beach face between about −1.5 m and the crest of beach face, which may have very different altitudes between 1 and 2.5 m. The hydrodynamic characteristic of beach waters is the development of swash processes, that back and forth of waters on the beach face, with very different morphometric and morphographic parameters, according to the degree of sea roughness. The effect of wave action upon the shore and beaches from the Western coast of the Black Sea differs a lot according to the direction under which the waves train moves forwards to the shore (Vespremeanu 1987):

- under the conditions of wind accelerations from North-East direction waves are formed which fall on water line at an angle of 130°–135° towards North, determining the refraction processes with the formation of longshore currents from North and to South, as well as the rip currents; therefore, the coastal alluvia drift is activated in the same direction, the losses of sands from the beaches being maximum;
- under the conditions of winds accelerations from South-East waves are formed which fall on the water line at an angle of 45° towards North, determining refraction processes with the formation of longshore currents and the alluvia drift from South to North, as well as the rip currents; therefore, the beaches are refilled with sand and extended; these reversible processes ensure the maintenance of shore structure and operation and of its subunits for long periods of time;
- the moderate—strong swell, with parallel waves at water line, leads to the movement of some large quantities of sand on the shore terraces (from depths of −10/−1.5 m) towards the summer berm, conditions under which the high part of the shore accumulates sand;
- the weak swell with parallel waves on the water line leads to the movement of sands from foreshores towards the beach face, with the development of the latter.

A specific distinct sector is the beach face on which the swash process is developed with the formation of beach current transporting the sand towards to the beach face crest, either directly, when the swash is imposed by the swell waves, either by lateral movements, in zig-zag, on the beach face, when the process is imposed by the waves transformed by refraction from a certain direction (Vespremeanu 1987).

The alternation of circulation from North and North-East (about 50% of the year) with that from South and South-East (about 35% of the year) is an essential process in the development of reversible processes on shores especially on beaches. If a single flow direction of coastal and shore currents would have prevailed, the low, sandy shores and the Romanian coastal beaches had been destroyed long time ago. The alternation of alluvia longitudinal drift driven by shore currents from the North sector with that from the Southern sector ensures the cyclic reinstatement of the alluvia stock from the transversal profile of the shore and, through this, could been maintained, under natural conditions, the shore and beaches at sizes according to the general sand stock from the shore (Vespremeanu 1987, 1989). This reversible character of shore processes is essential for a balance operation of the system (Vespremeanu 1987).

Human interventions, by the construction of perpendicular barriers on the water line which interrupted, to a large extent, the longitudinal flow of alluvia on shore, have disturbed the natural regime by generating erosions and sand storages under the barriers, preventing the development of natural reconstruction processes of beaches. The barriers having wide negative effects are those protecting the Sulina channel-branch as well as the protection barriers of Midia, Constanta and Mangalia harbors.

As shown above regarding the surface circulation from coastal regions, the following results:

- In the coastal waters from the North-West of the Black Sea there are no permanent currents.
- The dynamics of waters is closely related to winds, to the impact of Danubian waters and the morphography of coastal and submarine relief.
- At the exterior border (towards the deep basin) of the coastal waters, water jet enter detached from the Black Sea Rim Current, generating cyclone and anticyclone vortexes of small sizes without having a decisive influence over the surface circulation.
- The shore morphodynamics on the North-Western coast of the Black Sea is controlled by the *long shore drift* resulted from the refraction, defraction and reflection processes of waves approaching the shore from various directions.

Frontal processes in the Black Sea. The frontal circulation is represented by vertical currents from depth to surface, developed in the upwelling process both in coastal waters as well as in deep basin (Şerpoianu 1980; Bulgakov et al. 1999). The contrasts between the thermohaline features of coastal waters and those from the deep basin, as well as the interaction with the atmosphere generate forces that allow the formation of upwelling at the level of steep sector of pycnocline. These processes are intensive enough to transport from 200 to 300 m depth to the surface cold waters, rich in hydrogen sulphide.

The most frequent upwelling coastal processes occur on the Southern coast of Crimea, on the Romanian and Bulgarian coasts, especially at South of Constanta, as well as on the Anatolian coasts. The coastal upwelling from the Romanian sector is generated especially by winds from the South and South-West which determine a total Ekman transport towards East. Under these conditions, vertical current develop along the upper part of the continental slope and on the shelf which transport towards shore cold waters rich in H_2S and nutrients. The temperature drops of surface waters may reach gradients of 14–15 °C, sometimes higher (Şerpoianu 1980). The upwelling processes from the deep basin take place in the waters located above the continental slopes at West of Crimea and South of Kerch Strait (Staneva and Stanev 1998). In both cases, as in any marine area affected by upwelling, the primary productivity increases following the increase in nutrients quantity.

References

Aksu AE, Hiskott RN, Kaminski MA, Mudie PJ, Gillespie H, Abrajano T, Yaşar D (2002) Last glacial-holocene paleoceanography of the Black Sea and Marmara Sea: stabile isotopic, foraminiferal and coccolith evidence. Mar Geol 190:119–149

Algan O (2003) The connections between the Black Sea and Mediterranean during the last 30 Ky. 2003 Seattle annual meeting (November 2–5, 2003). Session No. 189: "Naoah's Flood" and the late quaternary geological and archeological history of the Black Sea and Adjacent Basins

Andrusov NI (1889) About Upper Pliocen deposits of the Cape Chauda on the Kerchenian peninsula. Proc Soc Nat Geol Mineral 20:11–12

Andrusov NI (1890) Predvaritelnîi octet ob uceastîi v Cernomorskoi globemernei expeditii 1890 gd.Izs.Russk.Geograf. Ov-va, 26, Moskva (in Russian)

Andrusov NI (1918) Geological structure of the Kerchenian strait bottom. Proc Russ Acad Sci 12 (6), 1:23–28

Antipa Gr (1910) Regiunea inundabilă a Dunării. Starea ei actuală şi mijloace de a o pune în valoare. Bucureşti: Instit.de Arte Grafice Carol Gobl

Antipa Gr (1937) Asupra curentului litoral NS din Marea Neagră. Commission europeenne du Danube, proces-Verbaux des Seances de la Session extraordinaires de Cannes 1936

Antipa Gr (1941) Marea Neagră. Vol. I, Oceanografia, Bionomia şi Biologia Generală a Mării Negre. Academia Română. Fondul Vasile Adamachi. Ed. Monitorul Oficial. Bucureşti

Arhanghelski AD (1927) Ob osadkov Cernovo Moria i ih zacenii v poznanii osadocinâh gorniah. Bull. Mosk. Obst. Isp. Pr., no. 3–4. (in Russian)

Arhanghelski AD, Strahov NM (1938) Geologiceskoe Stroenie I Istoria Razvitia Cernovo Moria. Moskva

Aubrey DG, Belberov Z, Bologa A, Eremeev V, Unluata U (1992) A coalition to diagnose the patient: comsblack and the Black Sea. Mar Technol 2(1):5–8

Bakan G, Büyükgüngör H (2000) The Black Sea. Mar Pollut Bull 41:1–6, 24–43

Banu AC (1961) Observaţii şi măsurători asupra oscilaţiilor de nivel actuale şi seculare ale apelor Mării Negre la ţărmul românesc. Hidrobiologia, Vol. II. Ed. Academiei Române

Bekers JM, Gregoire M, Nihoul JCJ, Stanev E, Staneva J, Lancelot C (2002) Modelling the Danube-influenced North-Western continental shelf of the Black Sea. I: hydrodynamical processes simulated by 3-D and box models. Estuar Coast Shelf Sci 54:453–472

Belyaev VI, Sovga EE, Lyubartseva SP (1997) Modelling the hydrogen sulphide zone of the Black Sea. Ecol Model 96:51–59

Berenbeim DIA (1959) Kerchian strait at Strabon time on the base of new data about Black Sea level variations. Covetskaia Arkheololghia, Nr. 4

Bogdanova AK (1969) Particularities de l'echange d'eau par le Bosphore. Rapp Int Mer Medit 19 (4):693–696

Boguslavsky SG, Zhorov VA, Porotskaya EM (1992) Vertical mixing in the Black Sea. In: Eremeev VN (ed) Complex ocenographic research. Utrecht, The Netherlands, pp 7–14

Bondar C (1972) Contribuţie la studiul hidraulic al ieşirii prin mare la gurile Dunării. Studii de Hidrologie, 19(2):23–28 (in Romanian)

Bondar C (ed) (1973) Marea Neagră în zona litoralului românesc. Monografie Hidrologică. Imh, Bucureşti, 520 Pag

Bondar C (1974) Zum findrigen des wassers des Scwarzen Meeres in die Donauarme, Kurzsreferate 17. Arbeitstagung de Internationalen Arbeitsgemeinshaft Donauforschung, pp 67–68

Bondar C (1986) Considerations on the water balance of the Black Sea. In: The Proceedings of the international meeting on the chemical and physical oceanography of the Black Sea. Report on the chemistry of sea water. University of Sweden, Göteborg

Bondar C, Filip I (1962) Contribuţii la studiul nivelurilor Mării Negre pe litoralul R.P.R. Studii de Hidrologie, Vol. III, Bucureşti (in Romanian)

Bondar C et al (1973) Le Mer Noire dans la zone du littoral roumain, *Monographie hydrologique*, Editura Technica, Bucharest, Romania, pp 516 (in French)

Brătescu C (1942) Oscilaţiile de nivel ale apelor şi bazinului Mării Negre în cuaternar. Bul. Soc. Rom. Regale de Geografie, Vol. LXI, 1–112 pp. citările sunt din aceeaşi lucrare apărută în volumul Constantin Brătescu, Opere Alese, Ed. Ştiinţifică, 1967, pp 193–278 (in Romanian)

Bulgakov SN, Kushnir VM, Martinez AZ (1999) Black Sea vertical circulation and extremes of the hydrochemical and hydrooptical parameters. Oceanological Acta

Buesseler KO, Livingston HD (1972) Times-series Profiles of 134Cs, 137 Cs and 90Sr in the Black Sea. In: Ozsor E, Mikaelyan A (eds) Sensitivity to exchange: Black Sea, Baltic Sea and North Sea, NATO ASI Series. Kluwer Academic Publishers

Cagatay MN (2003) Water exchange between Mediterranean and Black Sea during late Glacial-Holocene: evidence from Marmara And Black Sea. 2003 Seattle annual meeting (November 2–5 2003). Session No. 189: "Noah's Flood" and the late quaternary geological and archeological history of the Black Sea and adjacent basins

Caraivan G (1981) Evoluţia zonei Mamaia în cuaternarul târziu. Pontica XV:15–32

Caraivan Gl, Şelariu O (1985) Quaternary environmental changes of the Black Sea. Rev Roum Geol Geophys Geogr Serie De Geographie 29:13–20

Cotovu V (1958) Contribuţie la studiul efectelor dinamice ale mării asupra coastei româneşti şi alucrărilor de apărare. Hidrotehnica, Nr. 2. Bucureşti (in Romanian)

Cutter GA, Klukholm RS (1999) The cycling of particulate carbon, nitrogen, sulfur, and sulfur species (iron monosulfid, greigite, pyrite, and organic sulfur) in the water columns of Framvaren Fjord and Black Sea. Mar Chem 67:149–160

Dachev V (2000) Implications of Accelerated Sea-Level Rise (ASLR) for Bulgaria. Proceeding of survas expert workshop on European vulnerability and adaptation to impacts of accelerated sea-level rise (ASLR). Hamburg, Germany, 19–21 June 2000, pp 25–28

Degens ET, Paluska A (1979) Tectonic and climatic pulses recorded in quaternary sediments of the Caspian-Black Sea region. Sed Geol 23:149–163

Degens ET, Ross DA (1970) Oceanographic expedition in Black Sea. Die Naturwiss 7:349–353

Deuser WG (1971) Organic carbon budget of the Black Sea. Deep-Sea Res 18:995–1004

Ducet N, Le Traon PY, Gauzelin P (1999) Response of the Black Sea mean level to atmospheric pressure and wind forcing. J Mar Syst 22:311–327

Filipov DM (1965) The cold intermediate layer in the Black Sea. Oceanol 5:47–52
Flood RD, Ryan WBF, Wong HK, Giosan L, Panin N, Shimkus KM, Berne S, Dinu C (1999) Climate change and sedimentation in the Black Sea region in response to pleistocene climatic changes. ODP Preliminary Proposals 515. Received at http://www.msrc.sunysb.edu/Black_Sea/HTML/Future_2.html
Ginzburg AI, Kostianoy AG, Nezlin PN, Soliviev DM, Stanich-Ny SV (2002) Anticyclonic eddies in the northwestern Black Sea. J Mar Syst 32:91–106
Gokasan E, Demirbag E, Oktaz FY, Ecevitoglu B, Şimşek M, Yuce H (1997) On the origin of the Bosphorus. Mar Geol 140:183–199
Gregoire M, Stanev E (2001) General Presentation. J Mar Syst 31:1–2
Hartley Ch.A Sir (1862) Description of the Danube and of the works, recently executed at Sulina Mouth. Proc. Inst. of Civil Engineers, London
Hiskott RN, Aksu AE, Yaşar D, Kaminski MA, Mudie PJ, Kostylev VE, Macdonald JC, Işler FI, Lord AR (2002) Deltas south of the Bosphorus strait record persistent Black Sea outflow to the Marmara Sea since 10 Ka. Mar Geol 190:95–118
Huang Y, Freeman KH, Wilkin RT, Arthur MA, Jones AD (2000) Black Sea chemocline oscillations during the Holocene: molecular and isotopic studies of marginal sediments. Org Geochem 31:1525–1531
Karakas G, James A, Al-Barakati A (2002) Modelling subsurface dynamics in the Black Sea. Oceanol Acta 25:101–116
Konovalov SM (1995) Anthropogenic Impact and Ecosystems of the Black Sea. Bull. De L'institut Oceanographique, Monaco, No. Special 15:53–83
Konovalov SK, Murray JW (2001) Variations in the chemistry of the Black Sea on a time scale of decades (1960–1995). J Mar Syst 31:217–243
Kukushkin AS (1992) Some peculiarities of the Sea surface dynamics in the deep section of the Black Sea. Complex oceanographic research of the Black Sea. In: Eremeev VN (ed) Utrecht, The Netherlands
Kuprin PN, Sorokin VM, Dimitrov PC (1980a) Principles of classification and types of section of the upper quaternary deposits of the continental terraces. Geol Geophys Res Bulgarian Black Sea Zone 188–202
Kuprin PN, Kalinin V, Krustev TI, Morgunov YG (1980b) Basic features of the tectonic development of the continental terrace during the cainozoic. Geol Geophys Res Bul Black Sea Zone 124–140
Lein AYü, Ivanov MV (1991) On the sulfur and carbon balances in the Black Sea. In Izdar E, Murray JW (eds) Black Sea Oceanography. Kluwer Academic Publ., Dordrecht, The Netherlands, pp 307–318
Leonov AK (1960) Reghionalnaia Okeanografia. Leningrad
Lericolais G, Popescu I, Panin N, Guichard F, Popescu S (2003) Was the last rapid sea change in the Black Sea linked to a catastrophic event recordered by mankind? 2003 Seattle annual meeting (November 2–5 2003). Session No. 189: "Noah's Flood" and the late quaternary geological and archaeological history of the Black Sea and adjacent basins
Malciu V, Diaconu V (2000) Long-Term trend of sea level at the Romanian littoral. IOC Workshop Report No. 171, Annex III, pp 21–27
Mikhailova EN, Shapiro NB (1996) Modelling the transport and transformation of river waters on the northwestern shelf and in the deep Black Sea. Mar Hydrophys J 3:30–40
Muratov MV (1961) Istoricul cuaternar al bazinului Mării Negre şi compararea acestuia cu istoricul Mării Mediterane. Analele Româno-Sovietice, Seria Geologie-Geografie
Murray JW et al (1989) Unexpected changes in the oxic-anoxic interface of the Black Sea. Nat 338:411–413
Neretin LN, Volkov II, Bottcher ME, Grinenko VA (2001) A sulfur budget Black Sea anoxic zone. Deep-Sea Res I 48:2569–2593
Neuman GJW, Pierson W (1966) Principles of physical oceanography. Ed. Prentice-Hall., pp 181–184

Neveskaia LA (1965) Late quaternry bivalvia of the Black Sea, their taxonomy and ecology. Nauka, Moskva, 319 p (in Russian)

Oguz T, Aubrey DG, Latun VS, Demirov E, Koveşnikov L, Sur HI, Diaconu V, Besiktepe S, Duman M, Limeburner R, Eremeev V (1994) Mesoscale circulation and thermohaline structure of the Black Sea observed during hydroblack 91. Deep-Sea Res. Nr. 41(4):603–626

Oguz T, Besiktepe S (1999) Observations on the rim current structure, CIW formation and transport in the Western Black Sea. Deep-Sea Res I 46:1733–1753

Oguz T, Marray JW, Callahan AE (2001) Modeling redox cycling across the suboxic-anoxix interface zone in the Black Sea. Deep-Sea Res I 48:761–787

Oktay FY, Gokoşan E, Sakinc M, Yaltirak C, Imren C, Demirbag E (2002) The effects of the North Anatolian fault zone on the latest connection between Black Sea and Sea of Marmara. Mar Geol 190:367–382

Ostrovski AB, Izmailov LA, Balabanov IA (1977) Nov\Ie dannâie o paleogidrologiceskom rejime ciornovo moria v verhnem pleistoţene i goloţene. In: Paleogeografia I Otlojenia Pleistoţena Iujnâh Morei SSSR. Nauka, Moskow, pp 131–140 (in Russian)

Ovchinnikov IM, Popov YI (1987) Evolution of the cold intermediate layer in the Black Sea. Oceanol 27:555–560

Ozsoy E, Beşiktepe S (1991) Sources of double diffusive convection and impacts on mixing in the Black Sea. In: Brandt A, Fernando HJS (eds) Double-Diffusive Convection. American Geophysical Union

Ozsoy E, Unluata U (1997) Oceanography of the Black Sea: a review of some recent results. Earth-Sci Rev 42:231–272

Pfannenstiel M (1944) Diluviale Geologie Des Mittelmeergebietes, Die Diluvialen Entwiklundgstadien Und Die Urgeschichte Von Dardanellen, Marmara Meer Und Bosphorus. Geol Rundsch 34:334–342

Pirazolli PA (1996) Sea level changes: the last 20.000 years. Ed. Wiley

Preisinger A, Aslanian S (2003) The Black Sea during the last 20,000 years: sea level, salinity and climate. 2003 Seattle annual meeting (November 2–5, 2003). Session No. 189: "Noah's Flood" and the late quaternary geological and archaelogical history of the Black Sea and adjacent basins

Repina IA (1996) History of North-West part of Black Sea beach. In: Proceedings of the international workshop on MED and Black Sea ICZM, Sarigerme, Turkey, pp 357–364

Ross DA, Degens ET (1974) Recent sediments of Black Sea. The Black Sea Geol Chem Biol AAPG Memoir 20:183–199

Ryan WBF, Pitman III W (1998) Noah's flood. The new scientific discoveries about the event that changed history. Simon and Schuster, New York

Ryan et al (1997) An abrupt drowning of the black sea shelf at 7,5 Kyr BP. Geo-Eco-Marina, Vol. II, pp 115–126, Bucureşti-Constanţa.

Simonov AI, Altman EN (1991) Hydrometeorology and hydrochemistry of the USSR Seas. The Black Sea. Gidrometeoizdat Iv:430 p

Stanev EV, Beckers JM (1999) Numerical Simulation of seasonal and interannual variability of the Black Sea thermohaline circulation. J Mar Syst 22:241–267

Stanev EV, Staneva JV (2000) The impact of baroclinic eddies and basin oscillations on the transitions between different quasi-stabile states of the Black Sea circulation. J Mar Syst 24:3–26

Stanev EV, Beckers JM, Lancelort C, Staneva JV, Le Traon PY, Peneva EI, Gregoire M (2002) Coastal-open ocean exchange in the Black Sea: observations and modelling. Estuar Coast Shelf Sci 54:601–620

Staneva JV, Stanev EV (1998) Oceanic response to atmospheric forcing derived from different climatic data sets. Intercomparation study for the Black Sea. Oceanol Acta 21(3):393–417

Staneva JV, Dietrich DE, Stanev EV, Bowman MJ (2001) Rim current and coastal eddy mechanisms in an eddy-resolving Black Sea general circulation model. J Mar Syst 31:137–157

Sur HI, Ilyn P (1997) Evolutiona of satellite derived mesoscale thermal patterns in the Black Sea. Prog Oceanogr 39:109–151

Şerpoianu G (1973) Le bilan hydrologique de la Mer Noire. Cercetări Mar 5–6:145–153

Şerpoianu G (1980) Observations sur le phenomene d'upwelling au littoral roumain de la Mer Noire. Cercetări Mar 13:7–15

Şerpoianu G (1981) Observations on the phenomenon of upwelling near the Romanian Coast of the Black Sea. Rep Int Comm Mediterr Sea 27:13–24

Şerpoianu G, Nae I (1981) Observations saisonnieres sur le regime thermique des eaux marines sur le plateau continental roumain de la Mer Noire. Cercetări Mar 15:7–25, Constanţa

Skopintsev BA (1975) Formation of contemporary chemical composition of the Black Sea waters. Hidrometeoizdat, Leningrad, pp 366 (in Russian)

Soliatkin EV (1963) O vodom balanse Cernovo Moria. Okeanologhia, tom 3(6):986–993. Moskva. (in Russian)

Sorokin Yü I (1982) The Black Sea: nature, resources. Ed. Nauka, Moskwa, pp 216 (in Russian)

Spindler IB (1890) O gidrologiceskieh isledovanie Cernomorskoi expediţii 1890 g. Izv. Rusk. Geograf. Ob-Va 26 (in Russian)

Spindler IB, Wrangel FF (1895) Materialîi po ghidrologhii Cernovo i Azovskovo morei, sobravîie v expediţia 1890 i 1891 gg. Zap. Po Gidrogr. 5:20 (in Russian)

Tsatskin A, Heller F, Gendler TS, Virina EI, Spassov S, Pasquier JDu, Hus J, Hailwood EA, Bagin VI, Faustov SS (2001) A new scheme of terrestrial paleoclimate evolution during the last 1.5 ma in the Western Black Sea region: integration of soil studies and loess magnetism. Phys Chem Earth A 26(11–12):911–916

Tomalzin D (1985) Changing coastal oceanography of the Black Sea. I: North-Western Shelf. Prog Oceanogr 15:217–276

Unluata U, Oguz T, Latif MA, Ozosoy E (1990) On the physical oceanography of the Turkish Straits. In: Pratt LJ (ed) The physical oceanography of sea straits. Edit Kulver

Varuscenko SI, Varuscenko AN, Klige RK (1984) Izmenenie kaspicescovo moria i bestocinâh vodoemov v paleovremeni. Akademii Nauk Sssr. 290 p (in Russian)

Velichko AA (ed) (2002) Dynamics of terrestrial landscape components and inner marine basins of northern Eurasia during the last 130,000 years. GEOS 232 (in Russian)

Vespremeanu E (1986) Deltele Terrei. Terra 1:11–15, Bucureşti (in Romanian)

Vespremeanu E (1987) Probleme de Geomorfologie Marină. Universităţii din Bucureşti, p 128

Vespremeanu E (1989) Progrese în cunoaşterea reliefului litoral şi submarin cu privire specială asupra bazinului Mării Negre. Terra 3–4

Vespremeanu E (ed) (2003) Lucrările Staţiunii De Cercetări Marine Şi Fluviale Sfântu Gheorghe. Univercity Din Bucureşti, Bucharest, p 173

Vespremeanu E, Degeratu M, Bandoc G, Vespremeanu-Stroe A, Constantinescu St (2002–2005) Studiul interacţiunilor atmosferă-mare în mediul costier din România în vederea fundamentării ştiinţifice a managementului integrat al zonei de coastă. Grant CNCSIS. (in Romanian)

Winguth C, Wong HC, Panin N, Dinu C, Georgescu P, Ungureanu G, Krugliakov VV, Podshuveit V (1996) Upper quaternary water level history and sedimentation in the northwestern Black Sea. Mar Geol 167:127–146

Wong HK, Ludmann T, Panin N, Konerding P, Dinu C (2002) Northwestern lack sea: Upper quaternary water level and sedimentation. Mediterranean and Black Sea turbidity system and deep sea fans. CIESM Workshop Series No. 7, Bucharest, pp 85–89

Yilmaz Y (2003) Morphotectonic development of the southern Black Sea region and the surroundings. 2003 Seattle annual meeting (November 2–5, 2003), Session No. 189: "Noah Flood" and late quaternary geological and archaeological history of the Black Sea and adjacent basins

Chapter 6
The Sediments from the Black Sea

Abstract The sediment pattern in the nearshore zone of the Black Sea is governed by surface and longshore bottom currents and wave action (Ross et al. in Black sea sedimentary framework, 1978). In the deep basin, the sediment pattern is controlled by an isolated cyclonic current system and bottom morphology. Large quantities of detritus from the Danube, Dnieper, etc. are deposited and trapped on the broad western shelf, whereas the terrigenous material derived from the geosyncline drainage areas (Pontic and Caucasus mountains, and Krimean peninsula) easily crosses the narrow shelf and enters the deep basin, often in the form of turbidite deposits. Textural analyses of cores from the western and eastern basins reflect these differences in the shelf morphology. A rather uniform sedimentation pattern of mainly fine-grained material predominates in the western basin, whereas abundant turbidites and silty material in the cores off the eastern coast indicate high variability in the sedimentation pattern. Recent sedimentation in the Black Sea is governed by the deposition of terrigenous allochthonous material of low carbonate content and the autochthonous production of large quantities of biogenic carbonate material (coccolithophorids). The highest clay and carbonate content is in central areas of the western and eastern basins. Because the biogenic constituents are composed of clay-sized calcite, the total carbonate content, as well as the amount of the >2 μm fraction, increase simultaneously with the Coccolith portion.

Keywords Sediment pattern · Detritus · Terrigenous material · Coccolithophorids

The sediments from the Black Sea basin were studied even since the late 19th century when Andrusov (1890) makes the first descriptions of the sedimentary facies on the sea bottom. The researches from the interwar period were continued after the World War II and amplified after 1960. The expeditions on board of vessels Atlantis II (1969) and Glomar Challenger (1975), as well as the further researches bring novelties so that nowadays we have a clear picture over the sediments, their vertical stratification and areal distribution in the basin.

E. Vespremeanu and M. Golumbeanu, *The Black Sea*, Springer Geography,
https://doi.org/10.1007/978-3-319-70855-3_6

6.1 The Sediments on the Bottom of the Deep Basin

The shallow drillings allowed the identification on the bottom of the deep basin of three litological units deposited in the last 25,000 years (Ross and Degens 1974).

- The 30 cm thick Unit 1 consists of muds with coccolithophores rich in carbonates (60–80% $CaCO_3$) and low organic carbon (5–10%) deposited in the last 3000 years.
- The 40 cm thick Unit 2 consists of micro-laminated sapropel deposited during 7000–3000 B.P. Starting with 7090 + 180 years B.P. the quantity of organic carbon from sediments rise sharply to over 80% while the quantity of carbonates maintains itself at values of 15–20%.
- Unit 3 has a not specified thickness, continuing itself below the core level and consists of black laminated lutite alternating with coarser fractions, rich in iron sulfates and low in organic carbon (below 5%) and carbonates (4–5%).

The interface between Unit 2 and Unit 1 represents the fast shift from lacustrine regime existing in the Black Sea about 7200 years ago to the marine regime after the invasion of Mediterranean waters through Bosphorus Strait.

The three deep drillings performed during the Deep Sea Drilling Project expeditions on board of Glomar Challenger vessel in 1975 allowed the litological knowledge of the basin up to the maximum depth of 1075 m below the sea bottom level.

The S.380 location is the most significant. Here the drilling started from the bottom level—2000 m—and collected an undisturbed column of sediments of 1075 m length. Were identified 16 litological units with ages between Present Day and upper Miocene. The quaternary sediments are present in the first 650 m and represented by sapropels, terrigenous muds, chemical precipitates with the predominance of siderites ($FeCO_3$) or carbonates (calcite, aragonite, dolomite). Between 680 and 890 m pliocene sediments exist represented by carbonate chemical precipitates alternating with gravel and coarse sand, deposited under freshwater lacustrine conditions. The Upper Miocene sediments lay under 890 m depth on the drilling column and are represented by dolomite and black marl.

6.2 Sediments in the Present Day

The present day sediments are of terrigenous, biogenous and autogenous origin. *The terrigenous sediments* are transported by the Black Sea streams and rivers, as well as the strong winds which carry large quantities of dust from the pontic steppes. The smaller quantities of terrigenous sediments come from the abrasion of loess shores. The fluvial terrigenous sediments represent the most important source and the total quantity of detritic material transported from the Black Sea amounts to 149.45 million tonnes/year. The solid flows transported in suspension are distributed very unevenly: 83 million tonnes (55%) are transported by the Danube, 37 million tonnes (25%) come from the Caucasus mountains, 11 million tonnes from the pontic plateaus and plains and 17 million tonnes (11%) from the Pontic Mountains and Anatolia Plateau (Shimkus and Trimonis 1974). Most part of these sediments are transported in suspension during spring, except for the sediments transported by the Caucasus streams, which register maximum solid flows during summer. The streams and river also transport large quantities of ions found in the solutions. The annual chemical input flow in the Black Sea is of 80 million tonnes, of which 65% is attributed to the Danube. To this are added the mineral nutrients (nitrogen, ammonia, phosphates, silicones) with a total of 1.2 million tonnes (Shimkus and Trimonis 1974; Trimonis 1974; Ozsoy and Unluata 1997; Konovalov and Murray 2001).

The biogenous sediments come especially from the phytoplankton. The production of organic matter is of 812 million tonnes/year, which represents a lot comparing to other seas. The high values of the primary production is explained by the important input of nutrients, by the chemical exchanges between the anoxic layer and suboxic layer with the oxic layer, as well as through the low levels of salinity which favor the development of phytoplankton. The main groups presented in the phytoplankton are the diatoms, dinoflagellates and coccolitophores, all reaching the maximum of biomass during summer. Under these conditions, the average quantity of organic carbon is of 3 mg/l, twice larger than the average quantity from the World Ocean.

The autigenous sediments result from the chemical exchange processes developed in the biogenous sediments, below −1800 m, on the abyssal plain. Here, the pyrite type colloidal and meha-colloidal iron disulphides (Fe_2S) are continuously formed through the decomposition of the organic remaining under anaerobic conditions.

The sediment pattern in the nearshore zone of the Black Sea is governed by surface and longshore bottom currents and wave action (Ross et al. 1978). In the deep basin, the sediment pattern is controlled by an isolated cyclonic current system and bottom morphology. Large quantities of detritus from the Danube, Dnieper, etc. are deposited and trapped on the broad western shelf, whereas the terrigenous material derived from the geosyncline drainage areas (Pontic and Caucasus Mountains, and Krimean peninsula) easily crosses the narrow shelf and enters the deep basin, often in the form of turbidite deposits. Textural analyses of cores from the western and eastern basins reflect these differences in the shelf morphology. A rather uniform sedimentation pattern of mainly fine-grained material predominates in the western basin, whereas abundant turbidites and silty material in the cores off the eastern coast indicate high variability in the sedimentation pattern.

Recent sedimentation in the Black Sea is governed by the deposition of terrigenous allochthonous material of low carbonate content and the autochthonous production of large quantities of biogenic carbonate material (coccolithophorids). The highest clay and carbonate content is in central areas of the western and eastern basins. Because the biogenic constituents are composed of clay-sized calcite, the total carbonate content, as well as the amount of the >2 μm fraction, increase

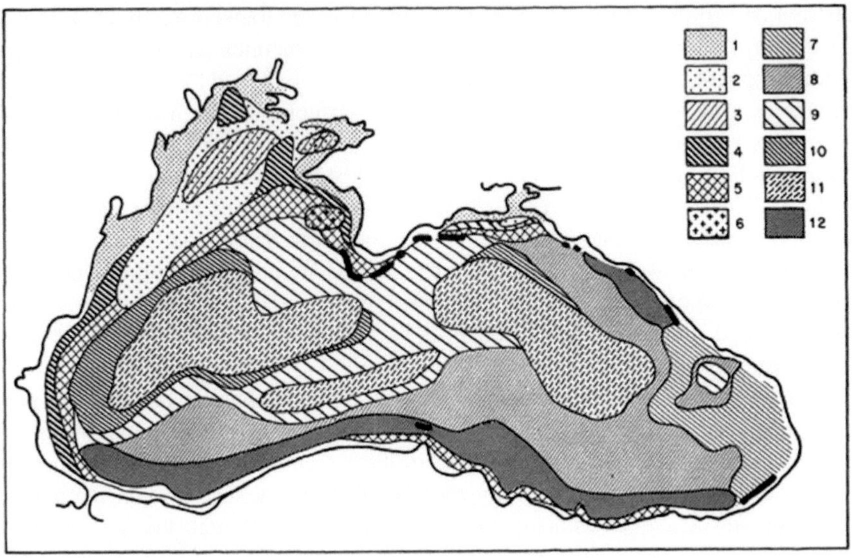

Fig. 6.1 Compositional-genetic types of modern Black Sea sediments, from Shimkus and Trimonis (1974). Shallow-water sediments: (1) organogenic-clastic, very coarse-grained and coarse-grained sediments; (2) carbonate-rich shelly sediments; ($CaCO_3$ 50%); (3) sediments covered by overgrowth of *Phyllophora*; (4) carbonate-poor and carbonate-bearing, organogenic-terrigenous mytilid muds ($CaCO_3$ = 10–50%); (5) carbonate-poor and carbonate-bearing phaseolina muds concretions; (6) deep-water sediments; (7) carbonate-free terrigenous sediments ($CaCO_3$ content 10%); (8) carbonate-poor organogenic-terrigenous muds ($CaCO_3$ content = 10–30%); (9) carbonate-poor, organogenic-terrigenous, finely dispersed muds ($CaCO_3$ content = 10–30%); (10) carbonate-bearing, organogenic-terrigenous, finely dispersed Coccolith muds ($CaCO_3$ content = 30–50%); (11) carbonate-rich (locally carbonate-bearing), finely dispersed Coccolith muds rich in organic matter; (12) modern sediments of considerable diversity with predominance of carbon-poor organogenic-terrigenic muds (*after* Ross et al. 1978). *Source* http://www.deepseadrilling.org/42_2/volume/dsdp42pt2_06.pdf

simultaneously with the Coccolith portion. On the basis of grain size, carbonate, and organic carbon content, 12 genetic types of modern Black Sea sediments can be distinguished (Fig. 6.1).

6.3 Distribution of Sediments

The distribution of sediments is influenced by the main dispersion factors: waves and currents. *In the coastal areas* (littoral and continental shelf), the waves and waves-generated currents, as well as the wind coastal currents are the main factors providing the dispersion of sediments. The littoral sediments are dominated by the terrigenous, terrigenous-biogenous or biogenous-terrigenous sands, according to the predominance of such fractions. The *Mya, Mytilus, Venus, Cardium* shells have a

10–60% contribution to the composition of shore and beach sand. The important input of rivers is stored mostly in deltas, lagoon and littoral fronts.

The continental shelf sediments differ a lot from one region to another. On the Danubian shelf the shell carbonate sediments prevail, with over 50% $CaCO_3$, mixed with clay minerals. On the Anatolian and Scythian self are muds with biogenous calcite mixed with biogenous aragonite and terrigenous carbonates. The Caucasian shelves are characterized by the presence of biogenous coccolith clay muds mixed with terrigenous calcite. The sediments on the continental slopes are differentiated according to the type and position of the slope.

On the Moessic and Scythian *continental slopes* terrigenous carbonate muds (10–15%) and mitilloide and phaseolonic muds are deposited. On the Caucasian continental slopes the clay muds are prevailing, to which are added the terrigenous muds low in carbonates. The Pontic continental slopes are covered by a large variety of sediments, with the predominance of the terrigenous muds low in carbonates mixed with coarse sediments.

The continental glacis sediments are represented by muds and grey clays low in carbonates and rich in clay minerals, as well as by greenish-grey organic-terrigenous muds.

On the abyssal plain there is an alternation of smooth coccolitic muds and greenish-grey silty clays rich in organic matter, iron sulphides and sapropel.

References

Andrusov NI (1890) Predvaritelnîi Otcet Ob Uceastîi V Cernomorskoi Glubomernei Expediţii 1890 Gd. Izs. Russk. Geograf. Ov-Va, 26. Moskva

Arhanghelski AD (1927) Ob Osadkov Cernovo Moria I Ih Zacenii V Poznanii Osadocinâh Gorniah. Bull. Mosk. Obst. Isp. Pr., Nr. 3–4 (in Russian)

Arhanghelski AD, Strahov NM (1938) Geologiceskoe Stroenie I Istoria Razvitia Cernovo Moria. Moskva

Konovalov SK, Murray JW (2001) Variations in the chemistry of the Black Sea on a time scale of decades (1960–1995). J Mar Syst 31:217–243

Ozsoy E, Unluata U (1997) Oceanography of the Black Sea: a review of some recent results. Earth Sci Rev 42:231–272

Ross DA, Degens ET (1974) Recent sediments of Black Sea. The Black Sea geology, chemistry, and biology. AAPG Memoir 20:183–199

Ross DA, Stoffers P, Trimonis ES (1978) Black Sea sedimentary framework. doi:10.2973/dsdp.proc.42-2.106, in DSDP Volume XLII Part 2

Shimkus KM, Trimonis ES (1974) Modern sedimentation in Black Sea. The Black Sea–geology, chemistry, and biology (pp 249–278). Memoir No. 20, Aapg

Trimonis ES (1974) Some characteristics of carbonate sedimentation in Black Sea. The Black Sea–geology, chemistry and biology (pp 279–295). Memoire No. 20

Chapter 7
Life in the Black Sea

Abstract The Black Sea is considered to be a huge laboratory naturally hosting oxic, hypoxic and anoxic water masses permanently existing due to strong vertical stratification. While strong vertical stratification supports isopycnal distribution of various biogeochemical species, the wide range of redox conditions supports specific processes rendering the Black Sea a unique place to study the Earth System responses to climate changes and anthropogenic forcing. Since a large part of the basin (i.e. approx. deeper than 100 m) is anoxic, life forms in the Black Sea display limited diversity and almost all pelagic and benthic fauna and flora dwell in the shallower upper oxic water layers. *The plants* are represented by over 304 species of macrophyte seaweeds, most of them red seaweeds, to which are added the brown and green seaweeds. The phanerograms are represented by 5 species only. *The animals* are represented by the majority of invertebrates groups, with a total of 1750 species, and the vertebrates include fishes, birds and marine mammals, with total of 164 species. The mammals are represented by two species of dolphins, the seal and the harbor porpoise: common dolphin (*Delphinus delphis* ssp. *ponticus*), common bottlenose dolphin (*Tursiops truncatus* ssp. *ponticus*), monk seal (*Monachus monachus*) and harbor porpoise (*Phocoena phocoena* ssp. *relicta*).

Keywords Biodiversity · Flora · Fauna · Fish · Mammals

7.1 Flora and Fauna of the Black Sea

The plants are represented by over 304 species of macrophyte seaweed, most of them red seaweeds, to which are added the brown and green seaweeds. The phanerograms are represented by 5 species only (Godeanu 1995).

The animals are represented by the majority of invertebrates groups, with an aggregate amount of 1750 species, and the vertebrates include fishes, birds and marine mammals, with a total of 164 species (Godeanu 1995).

E. Vespremeanu and M. Golumbeanu, *The Black Sea*, Springer Geography,
https://doi.org/10.1007/978-3-319-70855-3_7

7.2 Associations of Marine Organisms

The great associations of marine organisms are: plankton, nekton and benthos. The plankton consists of the total organisms which live by passive floating in the water mass, grouped in two categories: phytoplankton and zooplankton.

The phytoplankton is formed mostly of diatoms (136 species, representing about 80% of the phytoplankton composition), followed by the dinoflagellates (about 17%), cyanophyceae, chlorophyceae, silicoflagellates, in total 269 species of seaweeds (Celan 1989). The most frequent species of diatoms belong to the *Chaetouras* and *Coscinoides* genera.

The zooplankton is formed of 70 species, most of them tintinnids, rotifers and copepods, which prevail, as well as of cladocera, jellyfish and chaetognatha. The number of species forming the zooplankton is surprisingly low comparing to that from the Mediterranean Sea.

Of particular interest is the stenotherm plankton accommodated to the life in cold waters, which during winter and spring is present in the superficial waters, and during summer descends in the cold intermediate layer (CIL).

The nekton is formed of planktofags fishes such as the anchovy (*Engrauluis encrasicholus*), small horse mackerel (*Trachurus trachurus*), mullet (*Mugil sp.*), Atlantic mackerel (*Scomber scombrus*), and predatory fish like the Atlantic bonito (*Sarda sarda*), Mediterranean horse mackerel (*Trachurus mediterarraneus*) and benthofags, like snouted sole (*Solea nasuta*), flounder (*Pleuronectus flesus*), turbot (*Scophthalmus maeoticus*), sturgeons in the early stages.

The benthos is formed of 1790 species, most of them including polychaeta, nematodes, mollusks, crustaceans, bryozoas, echinoderms, tunicates, forming various biocoenosis up to the depth of almost 200 m (Băcescu and collaborators 1971).

7.3 Fishes, Birds and Mammals from the Black Sea

The fishes are represented by three species from the *Chondrichthyes* class and 164 species from the *Osteichthyes* class.

The Chondrichties class is represented by the piked dogfish (*Squalus acanthias*), thornback ray (*Raja clavata*) and the common stingray (*Dasyatis postinaca*). The dogfish is more widespread than people think; populating especially the shelf waters and the deep basin waters, at depths of 20–80 m, and frequently comes close to the shore. The most popular, by their economic value, are the anadromous migratory species: Atlantic mackerel, Atlantic bonito, horse mackerel, bluefish, anchovy and mullet. In the high seas are present the tuna (*Thunnus thynnus*) and swordfish (*Xifias gladius*).

The gobies, bleniidae, labridae, shallow-waters fish, live in the regions with rocky substrate, and the snouted sole, flounder, turbot, big-scale sand smelt, sturgeons live in the regions with sandy substrate.

The sturgeons (*Acipenseriformes* order) are fish with bone cartilaginous skeleton, present in the Black Sea coastal waters, from where they enter the rivers during the spawning period. The four species have a particular economic value, both for their tasty meat as well as for the caviar sought after in the fish markets: beluga (*Huso huso*), starry sturgeon (*Acipenser stellatus*), sterlet (*Acipenser rhutenus*) and Russian sturgeon (*Acipenser guldenstaedti*). In present, the populations of these species are increasingly reduced in terms of quantity due to the overfishing and the degradation of benthal biocoenosis where they are feeding.

A fish that appears temporary in the Black Sea is the eel (*Anguilla anguilla*), which reproduces in the Sargasso Sea in the Atlantic Ocean. The larvae and fry are transported by the Bay's Currents and the North Atlantic Drift for 2–3 years, reaching the high seas of Gibraltar Strait. Young eels (cives) pass, through Gibraltar Strait, in the Mediterranean Sea, entering on the streams flowing in it. Part of them arrives through the Aegean Sea and Marmara Sea in the Black Sea where they stay for a while in the littoral waters. From here they enter in Danube up to the upper basin, including on Romania streams, where they stay for 10–15 years, until sexual maturity, when they try to restore the way back to Sargasso Sea for spawning. Most of them die in the Mediterranean Sea and in the Atlantic waters in front of Gibraltar Strait. The population shall be restored by the eels that migrated on the American streams and come back, under proper conditions, in Sargasso Sea.

The typical marine birds are few and appear accidentally in the Black Sea space. This is the case of the manx shearwater (*Pufinus pufinus*), noticed above the coastal waters and in high seas especially during pre or post-nuptial migrations. The marine shore birds are much numerous, but they have a very large area of activity in the inland waters, often at large distances from the shore.

The mammals are represented by two species of dolphins, seal and porpoise.

The common dolphin (*Delphinus delphis* ssp. *ponticus*) surrounds the Pontic area, populating especially the waters of continental shelves. The maximum frequency is registered in the Crimea, Bulgaria and Turkey waters; sporadically it appears on other coasts. The effective number of dolphins from the Romanian waters was assessed at 600–800 individuals (*Black Sea Red Book Web Site*).

The common bottlenose dolphin (*Tursiops truncatus* ssp. *ponticus*) populates most part of the Black Sea waters but especially the high seas of Crimea, Caucasus and Anatolia coasts (*Black Sea Red Book Web Site*).

The monk seal (*Monachus monachus*) populates the waters of rocky coasts up to 50 m depth. The Mediterranean seal was frequent on the Anatolia and Crimea coasts, with sporadic presence in Bulgaria and Romania. In present, the population from the Black Sea is declining and strictly protected (*Black Sea Red Data Book*).

The porpoise (*Phocoena phocoena* ssp. *relicta*) populates the coastal waters surrounding the euxinic area, entering in Azov Sea, in the large lagoons and even on the rivers mouths.

7.4 Layering of Benthic Associations

The benthic associations are distributed on layers starting with the sea wall edge, located at 20–45 m, until the shelf edge located, under the Black Sea conditions, at 150–180 m depth.

The layer was defined as the benthic field sector, where the ecological conditions are characterized by some constant values or affected by regular variations. In present, 5 layers are accepted: supralittoral, mediolittoral, infralittoral and periazoic (Băcescu et al. 1971).

The supralittoral layer. The supralittoral layer was initially limited to the sector splashed by waves. We have extended this sector up to the upper limit of sea fronts due to the close, abiotic and biotic relations between the sea front, beach and water line. The sea wall flora is poor due to the adverse conditions. A few species of ruderal weeds are present, such as *Hordeum murinum, Malva pussilla, Atriplex tatarica.* At the sea wall base, especially on the wetter sectors, there is the association dominated by *Polygonum monspeliensis, Puccinelia distans* and *Spergularia marginata.* The sea wall fauna is formed of 32 species of birds, 6 species of reptiles and 3 species of mammals.

The beaches vegetation is differentiated according to the beach sizes. On high beaches associations with *Elymus sabulosus, Centaurea arenaria, Eringium maritimum, Convolvulus persicus, Medicago falcata, Salsola rutenica, Carex ligeria* develop. On dunes associations with *Elymus arenaria, Medicago falcata, Linaria genistifolia, Euphorbia suguieriana* develop, sometimes with bushes of *Crataegus monogyna, Tamarix pallasii, Salix fragilis, Hippophae rhamnoides*. The rocky supralittoral is represented by limestone outcrops from the base of the sea walls between Vama Veche and Midia Cape. The rock blocks and the tetrapods of harbor facilities and those from resorts fall into the same biotope.

The rocky supralittoral is an environment life characterized by the quasi-permanent waves splash. Under these conditions a multi-layer bioderma is formed (Băcescu et al. 1965): (i) at the upper part, less wetted, the cyanophyceae prevail; (ii) the middle part, very frequently wet by waves, is dominated by diatomaceous colonies; (iii) the lower part, permanently submerged, presents a rich and varied biocoenosis, with *Balanus*, cirripedia, crustacean, mysida, isopoda colonies. On the whole, the supralittoral is presenting itself as a complex biotic formation, strongly influenced by the sea. The winds and waves are adjustment factors and at a big storm most of the biocoenosis can be washed, which confers them a sharp dynamics.

The midlittoral layer. The mediolittoral layer is located between the beach face crest and the depth of 0.5 m, overlapping the beach face and foreshore, until the last ditch. From a hydrodynamic point of view corresponds to the waves breaking sector and swash sector. According to the nature of the substrate, two types are differentiated: the sandy midlittoral and rocky midlittoral (Băcescu et al. 1971).

Within the sandy midlittoral, there are two biocoenoses: biocoenosis of sands with average granulometry and sands with fine granumoletry (Băcescu and collaborators 1971). In the first one prevail the microbenthic species, including the *Donacilla cornea* seashell. The second biocoenosis is dominated by amphipods, in particular *Pontogammarus maeoticus*, which may reach densities of 50,000–60,000 specimens/m^2. The fauna of rocky midlittoral is very similar to that from the rocky supralittoral level. Here are formed well developed colonies of *Mytilaster lineatus*, and rarely *Mytillus galloprovincialis* and *Balanus improvisus.*

The infralittoral layer. The infralittoral layer is located on the shore terrace, between the extreme limit of the foreshore, at about −0.5 m and the edge of shore terrace, at −10/−12 m.

The sandy midlittoral is developed on the sequence of ditches and bars affected by the waves base. Within this level there are several biocoenosis dominated by mollusks (gastropods, tusk shells and bilvalvia) to which are added at least 100 of other psamobionte species (Băcescu et al. 1971). The rocky infralittoral, discontinuously present in South of Midia Cape, is domineated by gastropods and bivalvia, especially by *Mytillus galloprovincialis, Actinia equina, Mya arenaria, Rapana venosa*, the last two species being recently introduced in the Black Sea.

The circalittoral layer. The circalittoral layer is located between the shore terrace edge at about −10/−12 m and 100–120 m depth. In terms of geomorphology, this level is developed on the littoral slope, littoral glacis and the smoothest part of the continental shelf. Its fauna is poor and the predominant biocoenosis are those with *Mytilus galloprovincialis*, between 20 and 50 m depth, and those with *Modiolus phaseolinus,* between −50 and −100 m.

The periazoic layer. The periazoic layer overlaps the continental shelf edge, at depths of 100–180 m. Two biocoenosis are present here, the first with *Modiolus phaseolinus*, and second, at greater depths, with *Bugainvillia*, both poor in species. Below 200 m depth, the living creatures are missing, except for the sulphobacteria.

7.5 Geographical Origins of the Black Sea Fauna

The Black Sea fauna was formed during the basin's evolution, starting with the Sarmatian Sea until now. Therefore, in the actual fauna we find Sarmatian relict, Mediterranean specie, Atlantic species, Indo-Pacific species and freshwater species. From the total of 1650 species of invertebrates and vertebrates, most of them (1320 species) have Mediterranean origin, 305 species are freshwater and only 125 species are relict.

Sarmatian relict. Within the wide Sarmatian basin which was shared from the Vienna basins, Pannonic basin, Getic basin, Euxinic basin, over the actual plains of the Black Sea, through Manici tunnel, until the Caspian and Aral basins, a brackish fauna existed well conserved in the Caspian Sea, where it was protected against other marine influences. This fauna is present in the Black Sea as Sarmatian relict called also Ponto-Caspian relict. Amongst the invertebrates, this category includes gastropods, polychaeta, isopods, amphipods, mysis shrimp species. Amongst the vertebrates, the fishes are most numerous: most of sturgeon, mackerel and gobiidae species.

The Mediterranean origin fauna entered the Black Sea during at least three stages, the last one starting 7200 years ago. This category includes the majority of invertebrates and vertebrates species populating today the Black Sea waters.

The freshwater fauna includes euribionte species such as *Percaria demidoffi, Acipenser nudiventris* and *Acipenser guldenstadti*, which do fine at quite large variations of the salinity.

Recent invaders. Through various ways, but especially through the vessels' hulls and in the ballast waters of the oil tankers, 21 species of animals were accidentally introduced in the last 100 years (Muller1995; Gomoiu and Skolka 1996; Zaitsev and Mamaev 1997).

From the Atlantic Ocean come 7 species, of which 2 are crustacean cirripede (*Balanus improvisus* and *B. eburneus*), 2 are jelly fhish (*Blackfordia virginica* and *Bougainvilia megos*), 2 are mollusks (*Mya arenaria* and *Doridela obscura*) and a species of ctenophora (*Mnemiopsis leidy*). From the Pacific Ocean come the *Rapana thomasiana* gasteropod and *Cunearca cornea* bivalve. From the North Sea come 4 species, of which 2 are crabs species.

Most of these invading species accommodated well in the Black Sea waters, competing the local species and generating ecological imbalances (Muller 1995; Gomoiu et al. 2002).

Deliberately introduced species. For economic, medical and aquarium related purposes, 13 species were deliberately introduced in the Black Sea (Zaitsev and Mamaev 1997). For economic purposes, 2 species of crustacean were introduced, 1 species of bivalve and 9 species of fish. For medical purposes, it was introduced, from Italy the *Gambusia affinis* fish for fighting against the malaria transmitting mosquito, knowing that gambusia is a great consumer of mosquito larvae. For aquarium purposes, the *Oryzios latipes* was introduced from Japan.

References

Băcescu M, Gomoiu MT, Bodeanu N, Petran A, Muller G, Manea V (1965) Studii asupra vieţii marine în zona litorală nisipoasă la nord de Constanţa. ecologie marină, Vol. I. Ed. Academiei Române, pp 7–138 (in Romanian)

Băcescu M, Muller GI, Gomoiu MT (1971) Cercetări de ecologie bentală în Marea Neagră. analiza cantitativă, calitativă si comparată a faunei bentale pontice. ecologie marină, vol III. Ed. Academiei Române, pp 7–357 (in Romanian)

Celan M (1989) La flore algologique de la Mer Noire. Cercetări Marine, vol 22. IRCM, Constanta, pp 5–106

Godeanu SP (ed) (1995) Determinatorul ilustrat al florei si faunei Romaniei – Mediul marin. Ed. Bucura Mond 1:1–383 (in Romanian)

Gomoiu MT, Skolka M (1996) Changements Recents Dans La Biodiversite De La Mer Noire Dus Aux Immigrants. Geo-Eco-Marina, RCGGM 1:34–47

Gomoiu MT, Alexandrov B, Shadrin N, Zaitsev Y (2002) The Black Sea—a recipient, donor and transit area for alien species. In: Leppakoski E, Gollash S, Olenin S (eds) Invasive aquatic species in europe: distributions, impacts and management. Ed. Kulver, pp 341–350

Muller GI (ed) (1995) Determinatorul ilustrat al florei si faunei României, vol I. Mediul Marin. Ed. Bucura Mond, Bucureşti, 383 P

Zaitsev Y, Mamaev V (1997) Biological diversity in the Black Sea: main changes and ensuing conservation problems. In: Proceedings of the third international conference on the mediterranean coastal environment, Medcoast 97, Qavra, Malta, 11–14 Nov 1997, pp. 171–187

Chapter 8
Environmental Protection Issues

Abstract The activities of population bordering the Black Sea induce a huge anthropogenic pressure. Considering the entire surface of the Black Sea catchment area, the pressure is, of course, larger. This pressure is exercised by pollution, eutrophication, overfishing, endangerment of rare species, loss of biodiversity, introduction of new species, alteration of structure and operation of wetlands surrounding the sea, shores degradations, alteration of landscape aesthetic. Most of the coastal ecosystems were subject to significant changes. It is the case of benthonic ecosystem with *Phyllophora*, called the Zernov's field, located on the North-Western shelf, now destroyed almost in its entirety. Starting from the observation that the Black Sea condition is continuously worsening by the increase in pollution, nutrients enrichment, microbiologic contamination, introduction of exotic species, improper management of resources, irreversible loss of some habitats and serious peril of some ecosystems, a set of principles for action and political actions have been proposed (in the frame of the Black Sea Commission, EU Directives, cross-border cooperation etc.).

Keywords Anthropogenic pressures · Coastal ecosystem · Protected areas
Vulnerable ecosystem

8.1 International Perspective

In the last 30 years, the environment protection of the World Ocean became one of the fundamental global issues of the World. If in the '70s the protection of seas and oceans was hardly mentioned, after 1985, together with the understanding of the vulnerability of the great system, namely the World Ocean, the necessity of protection has a spectacular return in the main concerns for the faith of humanity in the future. The numerous international concerns for the marine environment protection in the '80s culminate with the discussion within the Global Environment Conference from Rio de Janeiro based on which the Agenda 21 was drafted, the fundamental document of this conference which embeds all the essential problems for the environment protection on Earth at all levels.

E. Vespremeanu and M. Golumbeanu, *The Black Sea*, Springer Geography,
https://doi.org/10.1007/978-3-319-70855-3_8

Chapter 17 of Agenda 21 has as subject: "*Protection of the oceans, all kinds of seas, including the enclosed and semi-enclosed seas, and coastal areas and the protection, rational use and development of their living resources*". The unprecedented development of the concerns on marine environment protection results also from the manifestations within the International Year of the Ocean.

The idea for organizing the International Year of the Ocean resulted from the discussions carried during the 17th session of the Intergovernmental Oceanographic Commission of UNESCO, from March 1993. The proposal launched by Portugal was reviewed and accepted by the General Conference of UNESCO from November 1993 as well as by the United Nations Economic and Social Council (ECOSOC) during the reunion from July 1994. Finally, the United Nations General Assembly adopts, by resolution A/RES/49/13 of December 1994, the declaration of 1998 as the International Year of the Ocean (IYO). The resolution was signed by 102 member states.

The main IYO objectives were to the raise the awareness of public attention, governments and decision-making institution about the importance of the ocean environments and the ways by which the marine resources may be sustainably used through an effective management.

The purposes of the activities from the International Year of the Ocean were oriented towards the large dissemination of efforts for understanding the following issues:

- The ocean and its vital resources for the life on Earth.
- The ocean affects our daily life, just like our daily life affects the ocean.
- The ocean has finite resources, our expectations being infinite.
- The ocean is the key of food, medications, energy and trade resources.
- The ocean is the next border with landscapes and life waiting to be discovered.

Therefore, IYO recommends:

- The ocean and coastal areas must be considered as having finite resources.
- The governments must cause priority actions for the purpose of protection and sustainable development in the oceanic and coastal environment.
- The joint global scientific efforts and the wide cooperation are the only ways to understand the structure and operation of the World Ocean. The results of scientific cooperation for knowledge about the ocean shall save many human lives and regional economies.

The important initiatives of IYO: Expo 98 from Lisbon, developed under the slogan "*Oceans, a legacy for future*", Charter of Oceans, oceanic education, development of scientific research, issue of postage stamps, cruises, publications.

The United Nations has declared 1998 the International Year of the Ocean as a celebration of this source of life and civilisation. But this international year is also a reminder of the need to protect this most precious of resources, an affirmation of our commitment to safeguard the rights of future generations, for whom we hold our planet—and its life-sustaining oceans—in trust.

The Charter of the World Ocean is another result of the fight for sea protection and was adopted by the international conference "*Summit of the Sea*" which took place in St. John, Newfoundland, Canada, in September 1997.

The Charter is the first international initiative for the coordination of the ocean and coastal environment protection, and was signed by most of the World's countries.

The need for protection and integrated coastal zone management is another issue concerning both the riparian states as well as the international institutions concerned about the global issues of the World.

The Coastal Environment is a system consisting of several sub-systems closely interconnected: relief, sediments, aquatory, atmosphere, ecosystems and social and economic subsystem. This very dynamic, fragile and vulnerable system lies over a water line length of over 500,000 km and an area of roughly 50,000,000 km^2 of the Earth area. Here are the large urban agglomerations of the planet and two thirds of the World population lives here. Therefore, the human pressure on this environment is very intense, manifesting itself by excessive pollution, extension of hyper-built spaces and destruction of ecosystems.

In the last 30 years it is noticed the intensification of coordinated attack of natural risk factors over the coastal environment on the whole and especially over the shore, which is the most dynamic and vulnerable part. The first attack comes from the sea and is due to the rising of waters level with rhythms ranging between +1.5 and +5 mm/year. The second attack is due to the increase in severe storms numbers. Both natural hazards are the results of the imbalances installed at the atmosphere-sea level interface due to global warming and pollution.

Following the human pressures and the actions of the natural hazards, the coastal environment is now subject to degradations which tend to become irreversible. Such degradations affect especially the ecosystems and the shore's relief. Annually, millions of hectares from the World Ocean shore are damaged or transformed, which represents priceless losses for the riparian states. Not only the economy is affected but also the population, human settlements, harbors and health resorts.

All these issues, as many others, imposed the development of concerns for the **coastal zone management**, which aims: "*solving the present and future issues from the coastal zone for ensuring a sustainable balance between the economic uses and a better quality of the environment. This aim may be achieved by using the deepened analysis of the natural processes and of the social and economic development*" (Coastline, EUCC, no. 3, 1993, p. 16).

8.2 Human Pressure on the Black Sea

The activities of population bordering the Black Sea induce a huge anthropogenic pressure. Considering the entire surface of the Black Sea catchment area, the pressure is, of course, larger. This pressure is exercised by pollution, eutrophication, overfishing, endangerment of rare species, loss of biodiversity, introduction of new species, alteration of structure and operation of wetlands surrounding the sea, shores degradations, alteration of landscape aesthetic.

Pollution

Heavy metals pollution. Heavy metals (mercury, cadmium, zinc, cooper, arsenic, nickel, chrom) are brought by rivers, contributing thus to waters pollution. Particularly worrying are the large quantities of cooper (about 1400–1500 tonnes/year), mercury (60 tonnes/year), cadmium 280 tonnes/year), zinc (6000 tonnes/year). Only the Danube transports annually in the Black Sea about 55 tonnes of mercury and 240 tonnes of cadmium (Konovalov 1995). These heavy metals are depositing in sediments, from where they are assimilated by mollusks and fishes and finally end in the human body and cause the most various diseases (Vespremeanu and Simion 2003).

Organochlorine pesticides pollution. The pesticides comes from agriculture and forestry, being transported by waters or wind, the maximum values being reached in the '60s and '70s, when large quantities of DDT and chlorinated hydrocarbon pesticides were used in an abusive manner. In 1989, the average concentration of DDT in the Black Sea was of 0.5–2 mg/l in high seas and 7–30 mg/l in coastal waters (Konovalov 1995). Now, the DDT quantity ended in the sea has dropped, but significant quantities of organochlorinated pesticide quantities are maintained (Europe's Environment 1995; Mee 2000).

Oil products pollution. The oil products end in the Black Sea water transported by Danube (about 53,000 tonnes/year), together with the wastewaters (about 30,000 tonnes/year) and from industrial sources (about 15,000 tonnes/year). An important contribution, difficult to be quantified, is due to the various types of vessels, especially oil tankers (Mee 2000).

Wastewaters pollution. Annually, a quantity of about 517 million m^3 of wastewaters are discharged in Black Sea, containing detergents, organic remaining, microbial germs. These waters, not treated or insufficiently treated, contribute to the eutrophication, the pollution with toxic substances and the microbial pollution of coastal waters.

Eutrophication

The first eutrophication phenomena occurred in the '60s (Gomoiu 1985). Gradually, the phenomena extended, generating the change in the phytoplankton structure and the ways of primary production.

Given that the quantity of nutrients was maintained at moderate values, according to the naturally controlled flows, the phytoplankton consisted of a large number of species, and the macrophytes algae were represented by 141 species (Muller 1995), including *Phyllophora* algae that formed broad fields on the North-Western shelf. The primary production of the phytoplankton and macrophytes algae provides the base of food chains ended with economically valuable fishes.

The forced development of agriculture in most of the countries bordering the Black Sea, but especially in the Danube, Dniester, Dnieper and Don basins by the excessive use of phosphatic and nitrogenous fertilizers has increased a hundred times the amount of nutrients in the coastal waters of the Black Sea (Mee 2000). Annually, roughly 160,000 tonnes of phosphorus and 800,000 tonnes of nitrogen ended in the Black Sea. The main sources of nitrogen are from agriculture (31%), wastewaters (26%), atmosphere (19%), industry (17%) and the main sources of phosphorus are from wastewaters (46%), agriculture (15%), industry (15%), atmosphere (8%) (Europe's Environment 1995).

Following the accelerated eutrophication the populations of some phytoplankton species have grown explosively, reaching enormous biomasses. The turbidity increased and the waters transparency dropped a lot. The amount of organic substance is much higher than the decomposition capacity and the surplus deposit on the sea bottom or beaches. Under these conditions, most of the benthic macrophytes have disappeared and the food chain which provided a balance productivity were seriously affected leading to a sharp decrease in the fishery production.

The protection against eutrophication may be achieved only by the adoption of modern strategies of rational use of chemical fertilizers, of an efficient treatment of the municipal wastewaters and of the resorts and the adoption of modern strategies for reinstatement of wetlands which have an essential filter role of the nutrients transported by rivers (Mee 2000).

Overfishing

The increasing number of fishing vessels (3000 fishing vessels in 1985, 3250 in 1990 and 4000 in 1997) and the use of harsh techniques (large size nets, but with small meshes, bottom trawling) led to the reduction in the effective populations of fish with economic value, especially the sturgeons and turbot. To this were added also the adverse effects of eutrophication as well as the warty comb jelly (*Mnemiopsis leidyi*) expansion, carnivore species feeding with zooplankton and fry.

For these reasons, the fishing status in the Black Sea, after the 1985, is particularly serious. Therefore, from about 850,000 tonnes of fished fish in 1985 in 1991 reached only 250,000 tonnes, and then in 1995 increased to 517,000 tonnes. Such increase was exclusively made based on the anchovy fished in the high seas of Turkish coasts, which illustrates the decline of fish species population with high economic value (Mee 2000).

At the same time, the number of industrially fished species with commercial value has been dramatically reduced, from 26 species, during 1960–1970 to only 5 species in present: anchovy (*Engraulis encrasicolus*), sprat (*Sprattus sprattus* and *Clupeonella cultiventris*), horse mackerel (*Trachurus mediterraneus* and *Trachurus trachurus*) (Bologa et al. 1995; Black Sea Region Homepage 2000).

The loss of biodiversity

The marine biology and ecology researches carried during 1890–1960, of which we highlight the particular contributions of Romanian specialists Gr. Antipa, I. Borcea and M. Băcescu, showed a biodiversity both at the ecosystems levels as well as at species level, which, even though did not reach the values from the Mediterranean Sea, was still remarkable.

After 1960 we witness the clear decline of some ecosystems and some species populations (Black Sea Red Data Book 1999; Black Sea Environmental Studies, UNDP 1999).

Most of the coastal ecosystems were subject to significant changes. It is the case of benthonic ecosystem with *Phyllophora*, called the field of Zernov, located on the North-Western shelf, now destroyed almost in its entirety.

The presence of *Rapana thomasiana* species in the Black Sea, originating from the Sea of Japan, which accommodated quickly, become a competitor of local mollusks species, led to the loss and disappearance of some mollusks population with consequences on the structure and operation of coastal ecosystems.

In a similar manner, the occurrence and large development of the *Mnemiopsis leidyi* population led to the dramatic loss of *Aurelia aurita* population and the zooplankton and, finally, to the reduction of planktivorous fish populations (Konovalov 1995).

The zooplaktonic communities lost a great part of their biodiversity in the period prior to the '60s. Therefore, the zooplankton biomass increased from 2.56 mg/m^3 in 1961, to 18.30 mg/m^3 in 1967, and to 155.56 mg/m^3 in 1976–1977, and reached 8719 mg/m^3 in 1983 (Bakan and Buyukgongur 2000). Such abnormal increase in biomass is accompanied by the loss of biodiversity, most of the zooplanktonic species populations have disappeared in favor of some species that became dominant. These are *Noctiluca miliaris*, with over 15,000 individuals/m^3, *Acartia clausi*, with over 5800 individuals/m^3, *Pleopsis polyphemoides*, with 1760 individuals/m^3. What is significant is the evolution of summer population biomass of *Noctiluca miliaris* species which increases from 200,000 tonnes, in 1950, to 3 million tonnes, in the '70s, and to 13 million tonnes, during 1980–1985 (Bakan and Buyukgongur 2000).

It is considered that nowadays 160 species are in danger, including also species which do not populate the Black Sea waters, being present only in the coastal areas (*Black Sea Red Data Book* 2000, Black Sea Environmental Programme 1993).

Alteration of wetlands structure and operation

The wetlands surrounding the Black Sea have a special role in filtering and partial neutralization of pollutants and fertilizers, contributing at the same time to the maintenance of biodiversity and natural balances. From this point of view, the deltas, firths and adjacent lagoons have an essential role.

The development of catchment areas and deltas led to the disappearance of some irreversible changes in the structure and operation of relative territorial systems, with the reduction of their favorable role over the coastal marine waters.

The need for protecting the Black Sea wetlands required the development of a reserve and harbors network as we shall see below.

The erosion of shores

Most of the Black Sea shores are affected by erosion processes which led to the annual loss of hundreds of hectares from the riparian states territories. The main affected areas are the low, sandy shores in the delta, lagoon and firth littorals.

The causes are multiple, but the main ones are the rise of sea level, the reduction in the amount of sandy alluvia transported by rivers, engineering coastal works which interrupt the alluvia flows along the shore.

The consequences are the induction of some risks on the entire Black Sea littoral, but in particular on the low shores, represented by loss of habitats, loss of biodiversity, impairment and damage of some resorts, human settlements, industrial and harbor objectives.

Alteration of landscapes aesthetic

The coastal marine landscapes had, in the early 20th century, an aesthetic value resulted from the prevalence of natural structures and the reduced human pressure.

The anthropic impact which developed quickly after the World War I, and in particular after 1950 led to the degradation of coastal landscapes by the growth of chaotic construction works (harbor facilities, resorts, cities, industries), without the concern for rational regional development and aesthetic. Parts of such human interventions are now abandoned and represent hard to cure scars which have a serious effect on the coastal landscapes aesthetic (Mee 2000).

8.3 Environmental Protection of the Black Sea

Amongst the numerous actions for environment protection in the Black Sea the main place is held by the Strategic Action Plan for Rehabilitation and Protection of the Black Sea, elaborated at Istanbul, Turkey, on October, 30th–31st 1996 and signed by the representatives of Bulgaria, Georgia, Romania, Russian Federation, Turkey and Ukraine governments.

The Strategic Action Plan is based on the spirit of United Nations Declaration on Rio de Janeiro Environment and on Agenda 21, as well as on the provisions of Bucharest Convention, April 1992 and Odessa Declaration, April 1993. The Contracting Parties first adopted the Strategic Action Plan for the Rehabilitation and Protection of the Black Sea (BS-SAP) in 1996, which was later amended in 2002. Further revisions to the BS-SAP were adopted by the Contracting Parties at the 2009 Ministerial/Diplomatic Conference in Sofia, Bulgaria.

Starting from the observation that the Black Sea condition is continuously worsen by the increase in pollution, nutrients enrichment, microbiologic contamination, introduction of exotic species, improper management of resources, irreversible loss of some habitats and serious peril of some ecosystems, a set of principles for action and political actions is proposed.

The principles for action are based on the sustainable development concept, the precautionary principle and principle of anticipation, use of clean technologies, principle of transparency and participation of civil society, as well as on the active collaboration between the riparian countries.

The politic actions consist of the reduction of pollution in the related catchment areas and Black Sea waters, management of living resources (use of commercial resources, protection of biodiversity, protection of habitats and landscapes), sustainable human development (environmental impact assessment, integrated coastal zone management, sustainable development of agriculture and tourism, involvement of civil society in decision-making processes).

Based on this strategic action plan, each riparian state drafted a national strategic action plan which contains also provisions on relevant international collaboration.

Fight against degradation processes and environment quality control

For the time being, the environment quality in the Black Sea is in a precarious condition induced by the increase of pollution, expansion and intensification of eutrophication, extension of harmful human activities on coasts, presence and expansion of some alien species, disappearance of some native species, presence of some major disruptions in the structure and operations of ecosystems, overfishing. Therefore, at international level as well as in the riparian countries, protection measures were taken aiming to the reduction of human pressure on all areas, improvement of some ecosystems status and elaboration of most adequate strategies for the rational use of resources and sustainable development, development of new education programs for environment protection, development of international cooperation.

At international level, one of the first actions took place at Bucharest and resulted in the "Convention for Black Sea protection against Pollution", known as the Convention of Bucharest. This convention contains three important protocols:

- Control of pollution source from land;
- Control of waste storage;
- Joint support actions in case of incidents.

It followed the "Ministerial Declaration on Black Sea protection" signed at Odessa on April, 7th 1993 and the "Strategic action plan for the Rehabilitation and Protection of the Black Sea", elaborated during the Conference of Istanbul, on October 30th–31st 1996. These documents were signed by all the riparian countries.

The main expected measures aim the reduction and monitoring of the pollution, management of living resources, sustainable regional development, elaboration of some national strategic plans for Black Sea environment protection, integrated coastal zone management, international cooperation between the riparian states and other states.

Safeguard of Black Sea wetlands

The protection of wetlands is partially provided for the time being by the establishment of a 28 protected areas system: Danube Delta Biosphere Reserve (Romania and Ukraine), Dniester Delta, Tiligulski firth, Tnedrovski and Iagorlkiski bays, Dnieper Delta, Djarilgah bay, Karkinit bay, Sivas lagoons, Molocinai firth, Obitocinaia, Bedrdiansk and Belosaraiskaia Peninsulas (Ukraine), Don Delta, Jejski firth, Bejsugski firth, Kuban Delta, Kazaltasski lagoon complex (Russia), Ciuria marshes from Kolhida (Georgia), Yesilirmak marshes, Kazalirmak Delta (Turkey), Sarikum marshes, Skarya Delta, Terkos bay and lagoons, Igneada Saka Longozu marshes (Turkey), Strangea marshes, Ropotamo marshes, Sabla Lake, Durankulak Lake (Bulgaria), Techirghiol Lake, Siutghiol Lake and Tasaul Lake (Romania) (Wilson and Moser 1994). Amongst these, the most important is the Danube Delta Biosphere Reserve.

Even since early 20th century, Gr. Antipa highlighted, in the memoires addressed to the Romanian Academy, the need to protect the unique vegetation and fauna in Europe. At the insistence of the great biologist, supported by E. Racoviţă, Tr. Săvulescu, Gh. Ionescu-Siseşti, Th. Buşnită and other famous scientists, the first Danube Delta reserve called "Pădurea Letea" is established in 1938. After the World War II the use of all Danube Delta resources is stringently imposed within the "Complex development plan of Danube Delta".

In 1965, the establishment of six reserves is proposed, with a total area of 41,064 ha and six refuges for nesting or migration. The reserves were marked by the National Commission of Natural Protected Areas within the Romanian Academy under the coordination of Prof. Val. Puşcariu and protected by the Resolution of Council of Ministers no. 891 from 1961. At the same time, the following species were declared as natural protected species and safeguarded as such, according to the law-decree no. 287/1950: common pelican, Dalmatian pelican, great egret, little egret, spoonbill, shelduck, ruddy shelduck, stilt, pied avocet. Law no. 9 of June, 22nd 1973, by art. 81 strengthens the status of these reservations and of natural protected areas.

In 1975 the new programme called "Complex valorisation of Danube Delta resources" was implemented (general coordinator Eng. Mircea Ardelan) which included also a project for the establishment of the Danube Delta National Park and the Razelm-Sinoe lagoon complex, with total area of 200,000 ha, of which 90,000 ha of integral reserves (Project Manager E. Vespremeanu, in collaboration with N. Botnariuc, Gr. Posea, N. Toniuc, Al. Filipascu). Unfortunately, the protection project was not applied anymore and in the late 1989 the general condition of Danube Delta nature protection was closer to the brink of disaster.

By the Government Decision no. 983 from 1990, by Law no. 82 of December, 7th 1993 and the Government Decision no. 248 of May, 27th 1994 the Danube Delta Biosphere Reserve was established.

Natural protected areas

A natural protected area is a land, underwater or underground area, with a legally set perimeter and a special protection and conservation regime, where plant and animal species, geographical, landscape, geological, paleontological, speological or any other type of feature or item, with a special ecological, scientific or cultural value, are present.

The aims and management regime of natural protected areas comprised in the national natural protected area network in Romania (as defined in compliance with the IUCN—International Union for the Conservation of Nature—terms of reference) are the following:

- **scientific reserves**: only scientific activities are allowed herein, with the consent of the competent scientific body. These reserves correspond to IUCN (International Union for the Conservation of Nature) Category I—Strict natural reserve: protected area managed mainly for scientific purposes;
- **national parks**: the management regime is set by in-house protection and conservation regulations and plans, approved by the entitled national scientific and administrative authorities. National parks correspond to IUCN Category II—National park: protected area managed mainly for ecosystem protection and leisure;
- **natural monuments**: the management thereof is made pursuant to a strict protection regime ensuring the conservation of specific natural features. Depending on their degree of vulnerability, people's access may be limited or not. Monuments of nature correspond to IUCN Category III—Natural monument: protected area managed mainly for the conservation of specific natural features;
- **natural reserves**: the management thereof is made in a differentiated manner, in respect to their characteristics, by active management measures to ensure the conservation of habitats and/or the protection of certain species, groups of species or biotic communities. Apart from scientific activities, as the case may be, organized tourism or education activities may be allowed. These reserves correspond to IUCN Category IV, namely management area of habitats/species (Fig. 8.1);
- **natural parks**: the management thereof aims at preserving the harmonious interaction between man and nature by protecting the habitat and landscape diversity, promoting the conservation of traditional land uses, encouraging and strengthening the traditional activities and practices of local populations. They correspond to IUCN Category V—Protected landscape: protected areas managed mainly for landscape conservation and leisure;
- **biosphere reserves**: the management thereof is made in compliance with in-house regulations and protection and conservation plans, pursuant to the recommendations of the Man-Biosphere Program, under the auspices of UNESCO (Fig. 8.1);
- **wetlands of international importance**: the management of these areas is aimed at their conservation and the sustainable use of the biological resources they

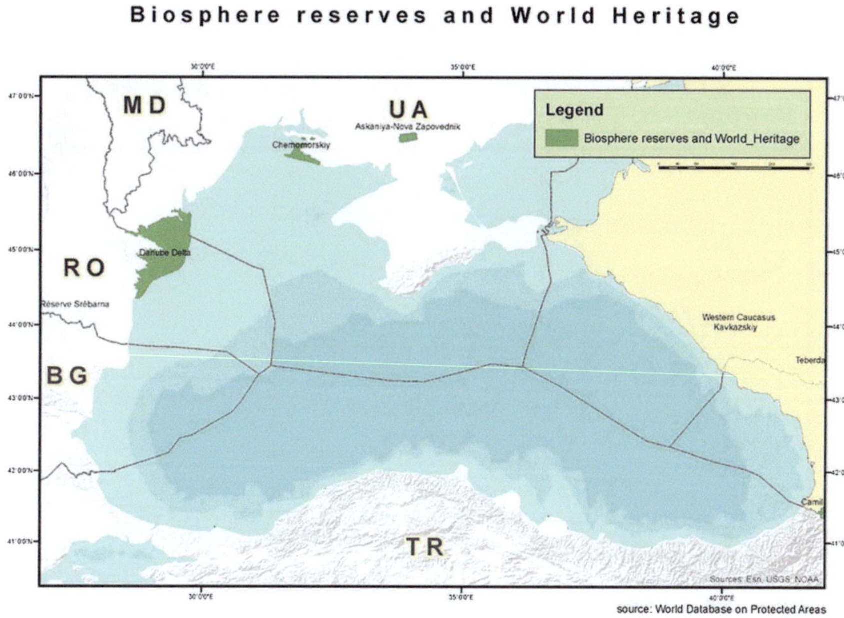

Fig. 8.1 Biosphere reserve and UNESCO world heritage sites at the Black Sea (Alina Spanu 2015)

generate, pursuant to the provisions of the Convention on the Conservation of Wetlands of International Importance, mainly as habitat of waterfowl (Fig. 8.2);

- **world heritage natural sites**: the management thereof is made pursuant to in-house regulations and protection and conservation plans, in compliance with the provisions of the Convention on the Protection of the World Cultural and Natural Heritage, under the auspices of UNESCO;
- **Special Areas for Conservation (SACs)**: designated for the conservation of natural habitats and habitats of species listed in Annexes 2, 3 and 4 of GEO no. 57/2007 on the regime of natural protected areas, the conservation of natural habitats, wild flora and fauna, as subsequently amended and supplemented. These special areas for conservation are included in the NATURA 2000 European network, after their status was acknowledged by the European Commission in compliance with the Habitats Directive (Fig. 8.2);
- **Special Protection Areas for birds (SPAs)**: are designated pursuant to the provisions of Directive 79/409/EEC on the conservation of wild birds and are included in the NATURA 2000 European network after their status was acknowledged by the European Commission in compliance with the Birds Directive (Fig. 8.2).

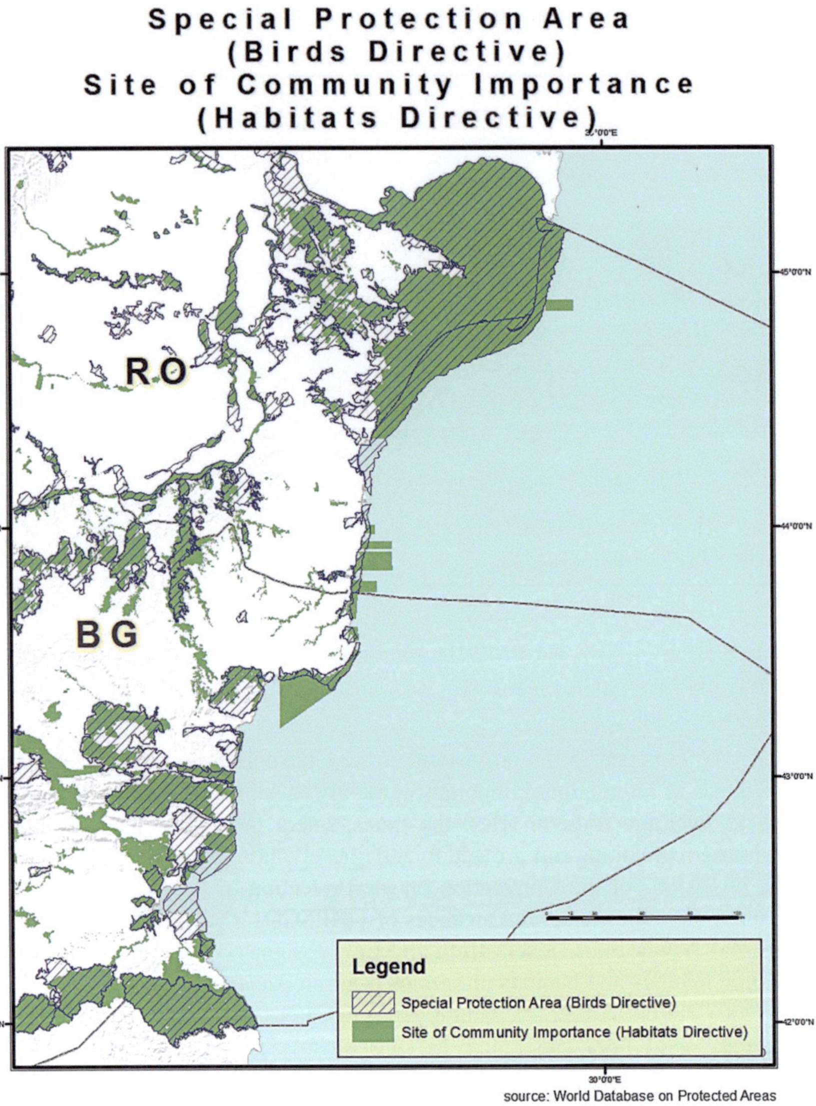

Fig. 8.2 Special protection areas for birds (SPAs) and sites of community importance (SCIs) in Romania and Bulgaria (EU Black Sea member states) (Alina Spanu 2015)

Protection of the most important coastal marine habitats

The hypoxia, anoxia and reduction of transparency processes due to eutrophication have determined the explosive development of the populations of some species detrimental to some other species, mass mortality, destruction of macrophytocenosis. To these is added the destruction of benthonic biocensis by trawling, deliberate or incidental introduction of exotic species, overfishing (Bakan and Buyukgungor 2000).

Amongst the most endangered biocoenoses are those dominated by mussel (*Mytilus galloprovincialis*), which includes over 100 species of invertebrates, field of brown algae (*Cystoseira barbata*), to which 60 species are associated, as well as the red algae fields (*Phyllophora nervosa*), with roughly 90 associated species (Zaitsev and Mamaev 1997; Bakan and Buyukgungor 2000).

One of the most interesting benthonic biocoenosis from the North-West of Black Sea is the one dominated by red algae *Phyllophora nervosa,* called Zernov's Field, after the name of Russian biologist Zernov who studied it in the early 20th century (Zernov 1913). The area of the *Phyllophora* field reduced continuously after 1960 following the reduction of transparency of waters and implicitly of the amount of light necessary for the development of benthonic macrophytes. Therefore, the area of *Phyllophora* field was of 10,000 km^2 in 1950, when the biomass was of 10 million tonnes. The field area reduced to only 50 km^2 in 1980, when the biomass was below 500,000 tonnes (Bologa et al. 1995). The destruction of most part of *Phyllophora* field led to the degradation of the entire biocoenosis and the disappearance of the majority of sponge, anemone, isopods, amphipods, crabs, squirts and over 40 fish species (Bologa et al. 1995).

Starting 2004 a trend of slow recovery of Zernov's *Phyllophora Field* is gradually recovering, but the balance between the species of Phyllophoraceae building the habitat has changed. *Phyllophora crispa*, which was dominant in the past, has become a secondary species, while now the dominant species is *Coccotylus truncatus*. The only species found in Romanian waters is *Coccotylus truncatus* (Fig. 8.3) (Micu et al. 2014).

Protection of rare species

As decided at the First Meeting of the Advisory Group on the Conservation of Biological Diversity of the Black Sea Commission, the Red Data List which covers all the threatened and rare species in the Black Sea ecosystem was compiled. It contains now 254 species (species which descriptions are still absent in the Black Sea Red Data Book are dimmed).

The list is divided into groups, a number of species in the list/book is given for each group: Plantae 63/43, Spongia 1/1, Polychaeta 2/2, Coelenterata 1/1, Insecta 7/4, Halacaridae 1/1, Crustacea 45/29, Mollusca 6/5, Echinodermata 2/1, Acrania 2/1, Pisces 52/41, Aves 66/25, Mammalia 6/6 (http://www.grid.unep.ch/bsein/redbook/index.html).

A number of 209 marine species are part of the Red List of marine endangered species at the Romanian littoral, elaborated in 2004 and revised in 2005 according to IUCN criteria (2004). Those species were distributed as follows: 19 macrophytes and vascular plants (9%), 45 invertebrates (22%), 141 fish (67%) and 4 mammals (2%). Six IUCN categories were applied for the 209 assessed species: Extinct—EX (6%), Endangered—EN (9%), Vulnerable—VU (5%), Near threatened—NT (13%), Least concerned—LC (16%) and Data deficient—DD (52%). More than half of all species assessed according to IUCN criteria were Data deficient (DD),

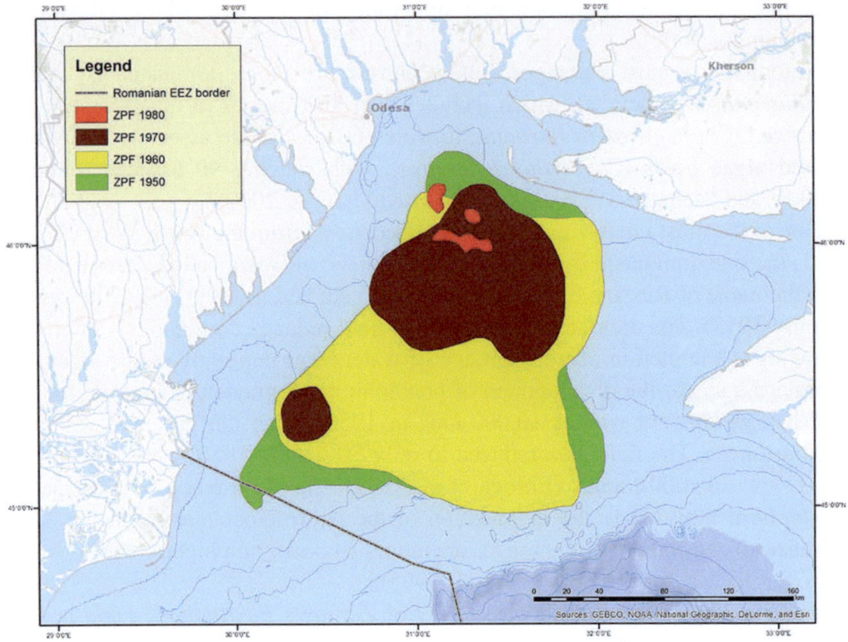

Fig. 8.3 Variation of Zernov *Phyloplora* field extent on the Northwestern shelf during the last decades (*Source* Micu et al. (2014), the recent recovery of Zernov's *Phyllophora* field extends into the Romanian Black Sea)

which means the necessity of further studies for clarification of their IUCN status. Out of 19 macrophytes and vascular plants species, 11 (57%) are EN, 2 (11%) are VU and 6 (32%) are DD. Among invertebrates, the 45 species from Red List are considered as follows: 13 (29%) EX, 3 (7%) EN, 6 (13%) VU, one (2%) LC, and 22 (49%) DD. Fish list contains 141 species, only two species (1%) being considerate as EN, two (1%) as VU, 28 species (20%) as NT, 31 species (22%) as LC and more than half—78 species (56%) DD. Regarding marine mammals, the three dolphin species are EN (Abaza et al. 2005).

Analyzing the evolution of biodiversity in the Romanian coastal waters, it is important to stress that after 1970, as result of the modifications of the marine environment, a general tendency of regress, dramatically in some cases has been registered, expressed by reduction of some planktonic invertebrate populations; disappearance of species, by mass mortalities among benthic populations; significant reduction of exploitable fish stocks and their diversity; reduction of the diversity of marine mammals. In the last 10 years, marine environmental conditions improved, leading to recovery of some populations, and to increasing of biodiversity, expressed not only by penetration of some alien species in the Black Sea, but also by reappearance of native species not encountered in the past 15 years.

In compliance with the Habitats Directive 92/43/EEC, the marine protected areas along the EU Black Sea coast (Romania and Bulgaria) provide the protection of (Zaharia et al. 2014):

- **Species**: *Tursiops truncatus, Phocoena phocoena, Delphinus delphis, Alosa immaculata, Alosa tanaica, Alosa maeotica, Huso huso, Acipenser gueldenstaedtii, Acipenser stellatus*;
- **Habitats**: 1110—Sandbanks which are slightly covered by sea water all the time, 1130—Estuaries, 1140—Sandflats and mudflats not covered at low tide, 1150*—Coastal lagoons, 1160—Large shallow inlets and bays, 1170—Reefs, 1180—Submarine structures made by leaking gases, 8330—Submerged or partially submerged sea caves.

Moreover, the Romanian and Bulgarian governments issued in 2006 a ban on the commercial fishing of all wild sturgeon species (including ship, Danube, stellate, sterlet and beluga sturgeon) for a ten-year period. The order also bans the trading of products or sub-products obtained from sturgeon captured in Romania, and requires any sturgeon captured accidentally to be released, regardless of its condition. The ban was ordered in response to concerns over the continuing decline of sturgeon populations and the extinction of other sturgeon species in Europe over the last century. This ban was extended in 2016 for a period of another 5 years, until 2021 (https://www.icpdr.org).

References

Abaza V, Boicenco L, Bologa AS, Dumitrache C, Moldoveanu M, Sburlea A, Staicu I, Timofte F (2005) Analysis of Hazardous substance contamination of the marine ecosystem in the Romanian Black Sea Coast, part of the marine strategy framework directive (2008/56/EEC) implementation, in Cercetări Marine/Recherches Marines nr. 36, ISSN: 0250-3069, pp 15–30

Abaza V, Alexandrov L, Bakan G, Boicenco L, Brasoveanu F, Coatu V, Dyakov O, Lazar L, Nenciu MI, Mateescu R, Nita V, Oros A, Spinu AD, Stegarescu V, Stoyanova AG, Tiganus D, Zaharia T (2015). In: Golumbeanu M, Nicolaev S (eds) (2015) Study on integrated coastal zone management. Ex Ponto Publishing House Constanta. Orginal English version, ISBN 978-606-598-397-7, 454 p

Bakan G, Buyukgungor (2000) The Black Sea. Mar Pollut Bull 41:1–6, 24–43

Black Sea Environmental Programme (1993) http://www.blacksea-commission.org/_tda2008.asp

Black Sea Red Data Book (2000) source http://eea.government.bg/eea/en/publicat/yearbook1/zpo/red-book.htm

Bologa AS, Bodeanu N, Petran A, Ţigănuş V, Zaitsev YP (1995) Major modifications of The Black Sea Benthic and planktonic biota in the last three decades. Bull. De L'institut Oceanographique, Monaco, No. Special 15, pp 85–110

Gomoiu MT (1985) Problemes Concernant L 'Eutrophisation Marine. Cercetări Marine, vol 18, IRCM, Constanţa

Konovalov SM (1995) Anthropogenic impact and ecosystems of The Black Sea. Bull. De L'institut Oceanographique, Monaco, No. Special 15, pp 53–83

Mee LD (2000) Cum să salvăm Marea Neagră. Un ghid al Planului Strategic de Acţiune Pentru Marea Neagră. Constanţa (in Romanian)

Micu D et al (2014) The recent recovery of Zernov's Phyllophora field extends into the Romanian Black Sea, 2nd General Assembly Coconet and Perseus, Athens, Greece, 27–30 Jan 2014

Muller GI (ed) (1995) Determinatorul Ilustrat Al Florei Şi Faunei României. vol I, Mediul Marin. Ed. Bucura Mond, Bucureşti, 383 p

Vespremeanu E, Simion B (2003) Oceanografie Medicală. Ed. Univ. Din Bucureşti, p 133

Wilson AM, Moser ME (1994) Conservation of Black Sea wetlands: a review and preliminary action plan. IRWB Publ. 33, p 76

Zaharia T, Maximov V, Radu G, Anton E, Spinu A, Nenciu M (2014) Reconciling fisheries and habitat protection in Romanian coastal marine protected areas. In: Leonart J, Maynou F (eds) The ecosystem approach to fisheries in the Mediterranean and Black Seas. Sci Mar 78S1:95–101. doi:http://dx.doi.org/10.3989/scimar.04028.25B

Zaitsev Y, Mamaev V (1997) Biological diversity in the Black Sea: main changes and ensuing conservation problems. In: Proceedings of the third international conference on the mediterranean coastal environment, Medcoast 97, 11–14 Nov 1997, Qavra, Malta, pp 171–187

Zerrnov SA (1913) K Vaprosu Ob Izucenii Jizni Cernovo Moria. Zap. Imp. Akademia Nauk, Seria Vii, 32, 1, pp 1–299

http://www.grid.unep.ch/bsein/redbook/index.html

https://www.icpdr.org

***Black Sea Environmental Studies, UNDP, 1999

***Black Sea Red Data Book Web Site, 1999, http://www.Grid.Unep.Ch

***Black Sea Region Homepage, 2000

***Coastline, EUCC, no. 3, 1993, p 16

***Europe's Environment, 1995

Chapter 9
Black Sea Coastal Population

Abstract The states bordering the Black Sea are Romania, Ukraine, Russian Federation, Georgia, Turkey and Bulgaria. The Black Sea regions have a permanent population of roughly 17.5 million inhabitants, to which are added 6–8 million tourists per year. This population is very unevenly distributed in the riparian states. The Black Sea catchment area has registered a total number of 63 harbors, as follows: Romania: 18 harbors (including rivers), Ukraine: 18, Bulgaria: 2, Moldova: 1, Turkey: 24 (Black and Marmara Sea) (Golumbeanu and Nicolaev in Study on integrated coastal zone management. Ex Ponto Publishing House Constanta, 2015). The important harbors on the Black Sea coast are: Constanta and Sulina (Romania), Odessa, Mariupol (Ukraine), Tuapse, Poti and Batumi (Georgia), Samsun and Istanbul (Turkey), Burgas and Varna (Bulgaria). In general tourism along the Black Sea coastal zones is rather well developed in Bulgaria, Romania and Ukraine. Besides Istanbul, Odessa and Varna are the largest cities within the Black Sea region, with a high capacity to attract tourists.

Keywords Riparian countries · Shipping routes · Harbors · Tourism

9.1 Population from the Riparian States

The states bordering the Black Sea are Bulgaria, Georgia, Russian Federation, Romania, Turkey and Ukraine which in 2004 had a total of 294,316,181 inhabitants (Table 9.1).

9.2 Population and Human Settlements from the Coastal Regions

The circumeuxinic regions have a permanent population of roughly 17.5 million inhabitants, to which are added 6–8 million tourists per year. This population is very unevenly distributed in the riparian states (Table 9.2).

E. Vespremeanu and M. Golumbeanu, *The Black Sea*, Springer Geography,
https://doi.org/10.1007/978-3-319-70855-3_9

Table 9.1 Population of the states bordering the Black Sea in the year 2004 (after Enciclopedia statelor lumii 2005)

No.	State	Number of inhabitants
1	Bulgaria	7,518,000
2	Georgia	4,693,000
3	Russian Federation	143,782,000
4	Romania	21,698,181
5	Turkey	68,893,000
6	Ukraine	47,732,000
	Total	294,316,181

Table 9.2 Population of the states bordering the Black Sea in the year 2012

Country	**Bulgaria**	**Georgia**	**Romania**	**Russia**	**Turkey**	**Ukraine**
Population (thousands, 2012)	8345	1889	9682	11177	8243	69,955

Source BSC (2013). Annual report of the Black Sea Commission Advisory Group on Development of Common Methodology for Integrated Coastal Zone Management (ICZM AG). Publications of the Commission on the Protection of the Black Sea Against Pollution (BSC), Istanbul, Turkey

The Ukrainian coasts, with an average population of 6.9 million inhabitants. The main coastal cities are: Odessa (1,115,000 inhabitants), Nikolaev (503,000 inhabitants), Kerson (355,000 inhabitants), Kerch (174,000 inhabitants), Eupatoria (108,000 inhabitants). Urban population amounts 66% of regional population. Only one city in the Black Sea region—Odessa—has population over 1 million inhabitants. Average population density in the region is 61.7 capita per km^2.

The Turkish coasts have a population of 8,243,000 inhabitants concentrated in the main cities: Kocaeli (1,339,000 inhabitants), Sakarya (815,000 inhabitants), Bolu (576,000 inhabitants), Zonguldak (647,000 inhabitants), Kastamonu (321,000 inhabitants), Sino (206,000 inhabitants), Samsun (1,199,000 inhabitants), Ordu (925,000 inhabitants), Giresun (457,000 inhabitants), Trabzon (757,000 inhabitants), Rize (256,000 inhabitants), Artvin (161,000 inhabitants). The Turkish Black Sea coast is comprised of 14 provinces: Artvin, Bartın, Düzce, Giresun, Kastamonu, Kırklareli, Kocaeli, Ordu, Rize, Sakarya, Samsun, Sinop, Trabzon and Zonguldak.

The Russian coasts have a stable population of 1,117,700 inhabitants in the following main cities: Novorosiisk (244,000 inhabitants), Soci (142,000 inhabitants), Tuapse (66,000 inhabitants), Anapa (58,000 inhabitants), Gelendgik (52,000 inhabitants).

The *Bulgarian coasts* have stable population of the amounts to a total of 834,500 inhabitants, the main cities are: Varna (340,000 inhabitants) and Burgas (211,000 inhabitants). 3 separate administrative districts bordering 16 municipalities, as follows:

- District of Dobrich: *Municipalities of Shabla, Kavarna and Balchik*;
- District of Varna: *Municipalities of Aksakovo, Varna, Devnya, Beloslav, Avren, Dolni Chiflik, Byala*;
- District of Burgas: *Municipalities of Nessebar, Pomorie, Burgas, Sozopol, Primorsko, Tsarevo*.

In *Georgia* the stable coastal population amounts a total of 650,000 inhabitants in the cities: Batumi (145,400 inhabitants), Poti (51,000 inhabitants), Kobuleti (34,000 inhabitants) (Fig. 9.1).

The Romanian coastal zone, bordering the Black Sea, is entirely located in the Dobrogea region, the latter covering an area of 15,485 km^2, corresponding to almost 6.5% of the entire Romanian territory and is defined within administrative boundaries of two counties Constanta and Tulcea. The coastal population from Romania amounts a total of 968,000 inhabitants, most of them in the cities of Constanta (350,000 inhabitants), Mangalia (48,000 inhabitants) and Sulina (4000 inhabitants) and the rest in the coastal rural settlements (Fig. 9.2).

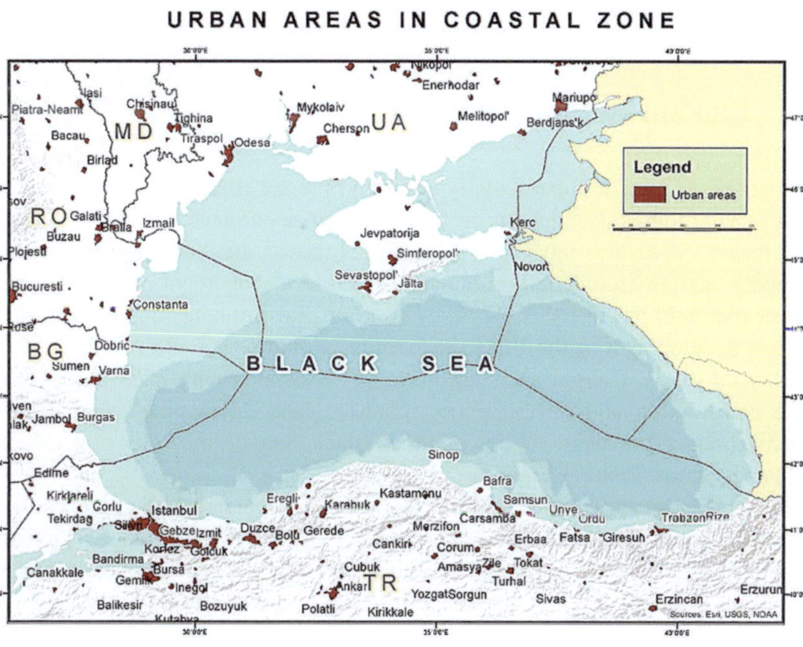

Fig. 9.1 Urban areas in the Black Sea coast (Golumbeanu and Nicolaev 2015)

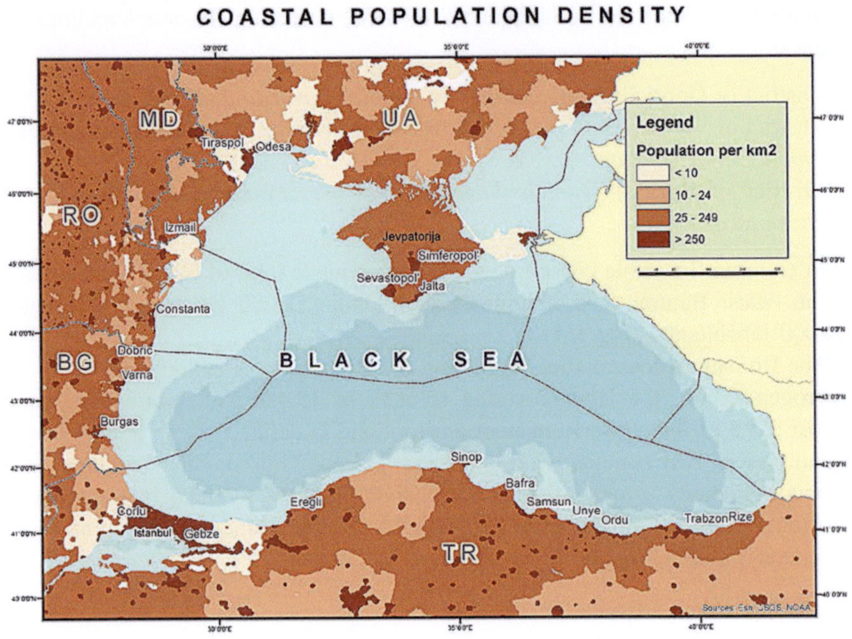

Fig. 9.2 Coastal population in the Black Sea coast (Golumbeanu and Nicolaev 2015)

9.3 Tourism and the Black Sea

Tourism development is particularly important for the Black Sea region where the most popular tourist destinations are the coastal areas, protected areas and historical settlements. Nowadays most of the Black Sea resorts are tackling different socio-economic and environmental problems. Therefore management of tourism growth is necessary and has to focus on policy measures for implementing the sustainable tourism development in the Black Sea region (Golumbeanu and Nicolaev 2015).

Coastal areas. Coastal areas are normally associated with mass tourism, large scale construction and infrastructure, intensive land development and extensive urbanization, a prevalent model in most Black Sea destinations (e.g. Mamaia, Romania, Sunny Beach Bulgaria).

Protected Areas. Tourism in protected areas is associated with appreciating and observing nature, scientific endeavour and education. This type of tourism is associated with minimal development of infrastructure and small scale interventions in areas of normally-strong control and restrictive management (e.g. Danube Delta Biosphere Reserve).

Rural areas. Tourism in rural areas covers a wide range of purposes (motivations) and is usually associated with visiting areas of special beauty, being in nature, low intensity activities but widely dispersed around low density-often remote- rural communities. In some areas agro-tourism falls within this category (e.g. Danube Delta villages).

Historical settlements and towns. Tourism is attracted to historic towns as a result of the built cultural heritage, urban amenities, lifestyle and cultural traditions, cultural events, etc. There can be several types of tourism in this category. The dominant type is mass tourism associated with large numbers of visitors centering on monuments, museums, etc. often of a short stay (even daily visits) (e.g. Nessebar, Bulgaria).

Tourism in the Black Sea region is an important industry. It benefits from general trends in world tourism, that have seen global tourism receipts grow an average of 8% per year from 1980 to 2000, while world economic growth averaged 3% (Lanza et al. 2005). Nonetheless, even at the national level, tourism in the Black Sea countries involves a relatively small number of visitors and expenditures. For example, Bulgaria, Russia, Turkey and Ukraine accounted for only 13.7% of international tourist arrivals in Europe in 2006. For receipts, the share is even smaller: the same four countries received only 8% of total European tourism receipts and only Turkey indicates a share of receipts greater than its share of international arrivals (4.1 and 4.5%, respectively) (Lanza et al. 2005).

Tourism, on the one hand, can (and today usually does) have major environmental impacts; on the other, it can be a way of promoting environmental awareness and encouraging natural regeneration.

In general tourism along the Black Sea coastal zones is rather well developed in Bulgaria, Romania and Ukraine (Fig. 9.3). Besides Istanbul, Odessa and Varna are

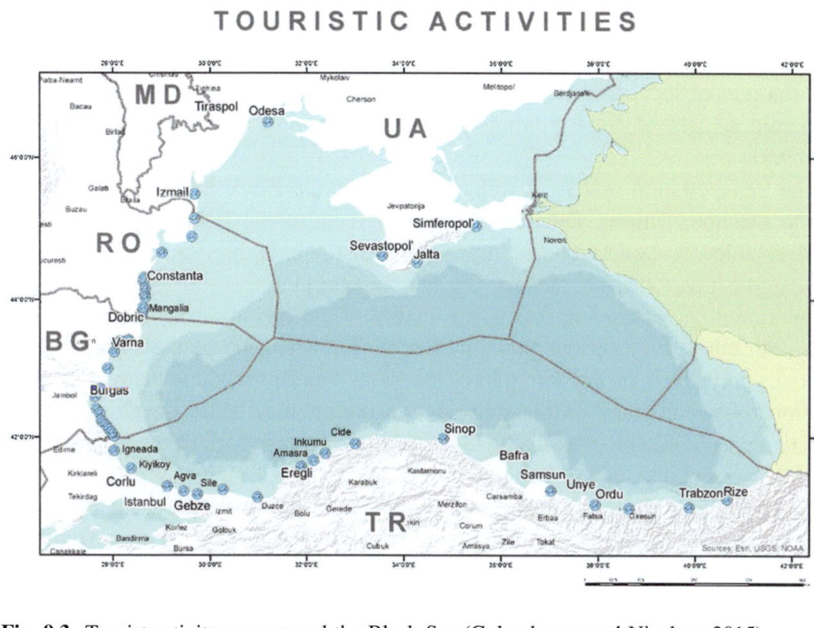

Fig. 9.3 Tourist activity area around the Black Sea (Golumbeanu and Nicolaev 2015)

the largest cities within the Black Sea region, with a high capacity to attract tourists. Tourism potential in Moldova and the hinterlands of Bulgaria, Romania and Ukraine could be better developed, put on an environmentally sustainable footing, and promoted to the local, regional and international tourism industry (such as travel offices and tourism agencies), by the tourism information providers, such as municipalities and regional authorities. The Black Sea countries mentioned are in possession of rich cultural heritage and provide beautiful green environment, mountains, cities, archaeological places, old monasteries, slumberous forests and endless hills. Moreover, the Bulgarian, Romanian and Ukrainian Black Sea coastal zone provide lot of opportunities for beach and water tourism. At the same time Black Sea municipalities suffer problems of city transportation, contamination by litter and wastes, bad water condition of coastal beach zone, lack of green belts and uncertainty of their legal status (Golumbeanu and Nicolaev 2015).

9.4 Key Harbors on the Black Sea Coast

It the post-war period, Black Sea traffic was dominated by the interests of the Soviet Union, which possessed nearly 50% of the Pontic coast. Turkey, a NATO member since 1952, holding one-third of it, targeted its traffic to its Aegean and Mediterranean ports (Urucu, 1999). The Soviet Union used to participate in world trade largely through the Black Sea harbours, through Odessa in the main. It also had strong trading ties, with Bulgaria in particular; sizeable cabotage traffic with the Black Sea ports of its own republics was also going on. The extension and equipment of its Pontic ports, as well as their modernization focused preferentially on Odessa or the military naval bases. The ports located on the eastern, Caucasian, side (largely of local importance at the time) would grow into tourist towns and economic centers of some administrative political units of the Caucasus region (Fig. 9.4).

The transportation system is well developed in the Black Sea Basin. Transport on the Danube, Dnipro, Dniester and Don Rivers to the Black and Azov Seas involves ships of the—river-sea‖ type. Sea ships include the ocean ships, such as dry cargo ships, and the tankers for the transportation of oil products. Water transportation adversely impacts water quality in the region during normal operations and represents a serious potential risk during accidents such as spills. Motor transport prevails in the western part of the Black Sea Basin where there is a highly developed road network, while railway transportation is better developed in the eastern region (Ukraine, Russia and Georgia). The extensive transportation network and intense mobility of the population and goods in the region affects the water quality negatively through such avenues as spills of oil products on the roadways and the use of inadequate technologies to treat wastewater coming from the industries servicing the transportation network (Golumbeanu and Nicolaev 2015).

With reference to naval transportation, the Black Sea catchment area has registered a total number of **63 harbors**, as follows: Romania: 18 harbors (including

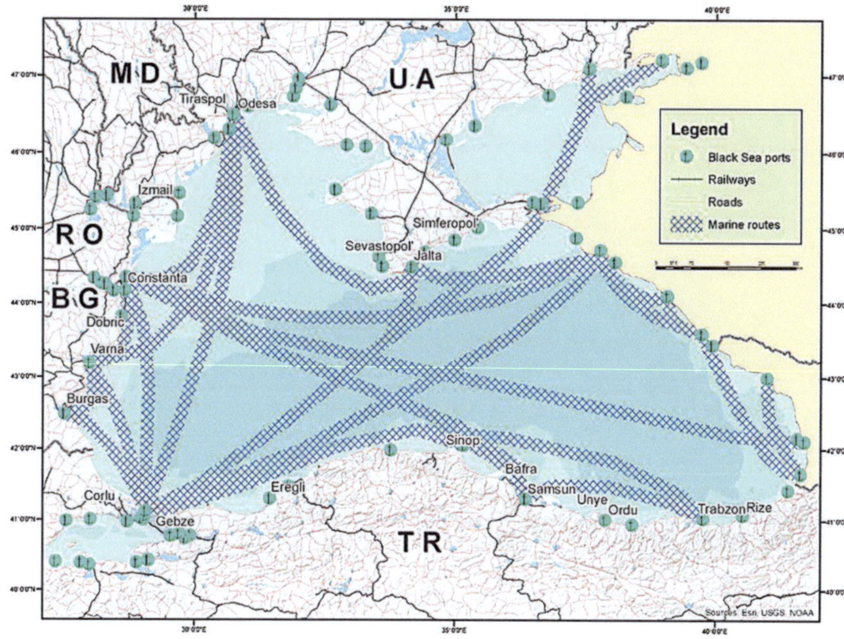

Fig. 9.4 Shipping routes in the Black Sea (Golumbeanu and Nicolaev 2015)

rivers), Ukraine: 18, Bulgaria: 2, Moldova: 1, Turkey: 24 (Black and Marmara Sea) (Golumbeanu and Nicolaev 2015). The important harbors on the Black Sea coast are: Constanta and Sulina (Romania), Odessa, Mariupol (Ukraine), Tuapse, Poti and Batumi (Georgia), Samsun and Istanbul (Turkey), Burgas and Varna (Bulgaria) (Fig. 9.5).

The largest of the Black Sea ports, Constanta, Romania, is located at the crossroads of the trade routes linking the markets of the landlocked European countries to Transcaucasus, Central Asia and the Far East. The port has excellent connections with the Central and Eastern European countries through the Corridor IV (rail and road), Corridor VII—Danube (inland waterway), to which it is linked by the Danube- Black Sea Canal, and Corridor IX (road), which passes through Bucharest. The two satellite ports Midia and Mangalia that are located not far from Constanta Port are part of the Romanian maritime port system under the coordination of Maritime Ports Administration SA Constanta. Constanta Port is both a maritime and a river port. Daily, more than 200 river vessels are in the port for cargo loading or unloading or waiting to be operated. Facilities offered by the port allow accommodation of any type of river vessel. The Port of Constanta is located on the Western coast of the Black Sea, at 179 Nm from the Bosphorus Strait and 85 Nm from the Sulina Branch, through which the Danube flows into the sea.

Fig. 9.5 Harbors around the Black Sea (*source* http://www.worldportsource.com/waterways/Black_Sea)

The connection of the port with the Danube River is made through the Danube-Black Sea Canal, which represents one of the main strengths of Constanta Port. Due to low costs and important cargo volumes that can be carried, the Danube is one of the most advantageous modes of transport, an efficient alternative to the European rail and road congested transport (Fig. 9.6).

After 1990, the Black Sea countries have been experiencing radical changes which affected also the volume and structure of the Pontic traffic. But for the Turkish ports, all the others have registered massive decreases in trading volumes, particularly those located on the west side of the sea, traffic volumes dropping sharply and suddenly within a span of several years. While the 1988–1989 Constanta harbor traffic turnover was 62 million tons, the steep decline in 1990 reduced it by half, a situation that continued for another three years.

After a first episode of decline caused by the confusion of transition, the port traffic of the former 'socialist' states recorded a timid comeback. The eastern and north-western ports appear to have fresh opportunities of development. For countries like Georgia and the Ukraine, which turned independent after 1991 when the Soviet Union fell apart, and even for Russia, these harbor have remained the principal, if not the only gateway to the word. They are striving hard to regain as much as possible of the role played in the Black Sea traffic during Soviet times. After 1990, apart from a decrease in traffic volume and structure due to the

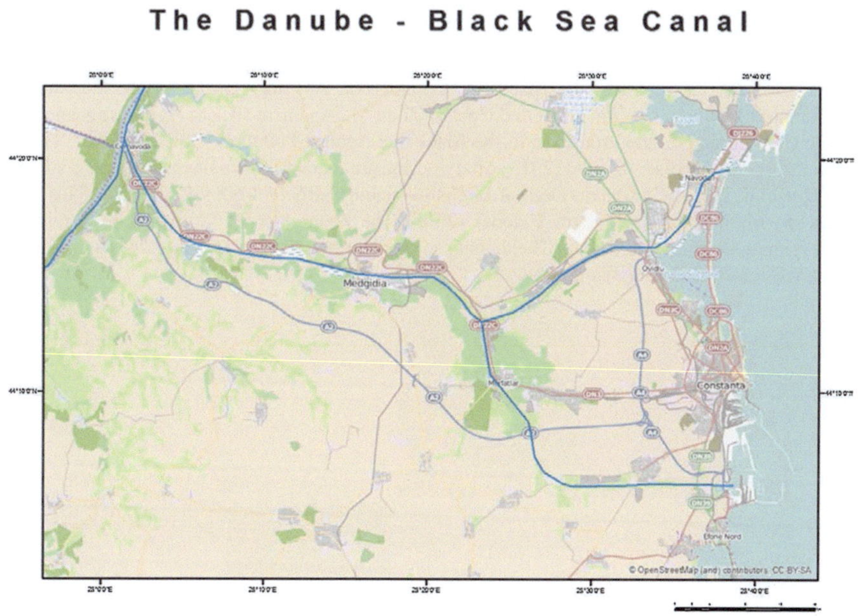

Fig. 9.6 Danube-Black Sea canal

economic difficulties of riverine countries in transition to the market economy, both the geographical area and the direction of shipments suffered radical changes. While cabotage traffic among the former Soviet ports decreased, the trade record of the eastern and southern ports registered increases. Profiting by the situation, Turkey's interest in the Black Sea ports was rekindled by a greater transit offer from the Middle East and the Persian Gulf area that combines land traffic up to Turkey's Pontic ports, with maritime transport on the Black Sea up to the European ports on the western and north-western side and farther on to Central and Western Europe. Bilateral agreements provide for a ferry-boat maritime line between Constanta and Samsun, but actual implementation depends on the two countries' goods offer. Besides, lots of taxes might diminish atractivity or make the projects uninteresting economically.

In a similar situation is the Constanta (Romania)—Poti line, which presents real advantages in connecting the Caucasian areas to the European Pontic zones. Port capacity is another factor discriminating the Pontic ports. While Constanta and Odessa have it in excess, the Caucasian ports lack much of it, so they cannot increase port traffic volume or diversify its structure. The new geopolitical situation seems to benefit the port of Novorossiysk, which is expected to become Russia's main Black Sea harbor.

References

BSC (2013) Annual report of the Black Sea Commission Advisory Group on Development of Common Methodology for Integrated Coastal Zone Management (ICZM AG). Publications of the Commission on the Protection of the Black Sea Against Pollution (BSC), Istanbul, Turkey

Golumbeanu M, Nicolaev S (eds) (2015) Study on integrated coastal zone management. Ex Ponto Publishing House Constanta. Orginal English version, ISBN 978-606-598-397-7, 454 p

Lanza A, Markandya A, Pigliaru F (eds) (2005) The economics of tourism and sustainable development. Edward Elgar, Cheltenham, UK, Northampton, MA. http://www.worldportsource.com/waterways/Black_Sea

Chapter 10
International Cooperation in the Black Sea Basin

Abstract Strategically placed between the European Union, candidate countries and neighbours, the Black Sea region constitutes an axis of increasing geopolitical importance for Europe. With its vast marine, energy and other natural resources, unique ecosystems and diverse interests ranging from shipping to tourism and fisheries to oil pipelines, nowhere is regional cooperation busier or more diverse. The governments of Black Sea countries are signatory parties to several international treaties relating to regional environmental cooperation. The significant role of the national governments in integrated approach of coastal zone should be realized through cooperation and partnership with other governments, the business sector and local communities (Golumbeanu and Nicolaev in Study on integrated coastal zone management. Ex Ponto Publishing House Constanta, 2015). Today, the *Convention on the Protection of the Black Sea Against Pollution* (also referred to as *Bucharest Convention*) is one of the most known European Regional Sea Conventions, establishing the legal ground for combating pollution from land-based sources and maritime transport, achieving sustainable management of marine living resources and pursuing sustainable human development in the Black Sea Region.

Keywords Regional cooperation · Riparian countries · International agreements

Strategically placed between the European Union, candidate countries and neighbours, the Black Sea region constitutes an axis of increasing geopolitical importance for Europe. With its vast marine, energy and other natural resources, unique ecosystems and diverse interests ranging from shipping to tourism and fisheries to oil pipelines, nowhere is regional cooperation busier or more diverse.

The governments of Black Sea countries are signatory parties to several international treaties relating to regional environmental cooperation. The significant role of the national governments in integrated approach of coastal zone should be realized through cooperation and partnership with other governments, the business sector and local communities (Golumbeanu and Nicolaev 2015).

E. Vespremeanu and M. Golumbeanu, *The Black Sea*, Springer Geography,
https://doi.org/10.1007/978-3-319-70855-3_10

10.1 The Bucharest Convention

Today, the *Convention on the Protection of the Black Sea Against Pollution* (also referred to as *Bucharest Convention*) is one of the most known European Regional Sea Conventions, establishing the legal ground for combating pollution from land-based sources and maritime transport, achieving sustainable management of marine living resources and pursuing sustainable human development in the Black Sea Region. The activities under the Convention allowed to significantly increase the public involvement, address transboundary environmental issues and to introduce sound environmental decision making related to the sustainable use of the Black Sea.

Since the signing in Bucharest in April 1992, and ratified by all six legislative assemblies of the Black Sea countries in the beginning of 1994, the Convention has become a powerful instrument of International Environmental Law in the Black Sea Basin. The implementation of the Convention is managed by the Commission for the Protection of the Black Sea Against Pollution and its Permanent Secretariat in Istanbul, Turkey. The Black Sea Commission comprises one representative of each of the Contracting Parties (Bulgaria, Georgia, Romania, Russian Federation, Turkey and Ukraine) to the Bucharest Convention.

Strategic Action Plan for the Rehabilitation and Protection of the Black Sea (BS-SAP). The Contracting Parties first adopted the Strategic Action Plan for the Rehabilitation and Protection of the Black Sea (BS-SAP) in 1996, which was later amended in 2002. Further revisions to the BS-SAP were adopted by the Contracting Parties at the 2009 Ministerial/Diplomatic Conference in Sofia, Bulgaria.

In the BS-SAP (both the 1996 and 2009 versions), the Contracting Parties elaborated on the goals and objectives that were laid out in the Bucharest Convention. Much of the BS-SAP was focused on how best to establish working relationships between the national frameworks of the Contracting Parties, outside states, and other groups that would be important in achieving the goals set out in the Bucharest Convention. The 2009 BS-SAP also formulated clear ecosystem quality objectives, corresponding targets (short-, mid- and long-term) to be achieved, and indicators of success.

The 1996 BS-SAP obligated the Commission to initially establish seven Advisory Groups ("AGs") and seven Regional Activity Centers ("RACs") that would focus on topics deemed to be a priority under the Bucharest Convention and the 1996 BS-SAP.

The *Advisory Groups to the Black Sea Commission* are its main source of expertise, information and support to in implementation of the *Black Sea Strategic Action Plan (BS-SAP).* There are seven Advisory Groups, as follows:

- ESAS—Advisory Group on the Environmental Safety Aspects of Shipping
- PMA—Advisory Group on the Pollution Monitoring and Assessment
- LBS—Advisory Group on Control of Pollution from Land Based Sources
- IDE—Advisory Group on Information and Data Exchange

- ICZM—Advisory Group on the Development of Common Methodologies for Integrated Coastal Zone Management
- CBD—Advisory Group on the Conservation of Biological Diversity
- FOMLR—Advisory Group on the Environmental Aspects of the Management of Fisheries and other Marine Living Resources

The Activity Centers support the activities of the Black Sea Commission's Advisory Groups. They are defined in the above mentioned seven strategic sectors.

10.1.1 Cooperation under the Black Sea Commission

The 1996 BS-SAP also called for the Commission and all of the Contracting Parties individually to encourage and pursue coordination among various regional bodies, non-governmental organizations (NGOs), the United Nations and other agencies regarding the sustainable development of the Black Sea region. One group identified in the BS-SAPs is the Organization of Black Sea Economic Cooperation (BSEC). BSEC includes all of the Contracting Parties of the Bucharest Convention, as well as many other non-coastal states from the Black Sea region. BSEC was created to increase economic cooperation among the states of the wider Black Sea region. In addition to the BS-SAPs, a joint task force (known as the DABLAS task force) was established in 2001 with the specific goal of coordinating the protection of the water and water-related ecosystems of the Danube River Basin and the Black Sea Basin (as the Danube empties into the Black Sea). The Commission, with the help of the United Nations Environment Programme (UNEP) and the United Nations Development Programme (UNDP)/Global Environment Facility (GEF), has also drafted a Memorandum of Understanding with the International Commission for the Protection of the Danube River (ICPDR) to agree on common goals for protection of the two water systems.

In 2008, the Ministers of Foreign Affairs of the EU countries and the countries of the wider Black Sea region issued a joint statement to initiate the Black Sea Synergy cooperation. The *Black Sea Synergy* is intended to encourage greater involvement by the EU in the Black Sea and to increase regional cooperation.

Furthermore, there were also a number of foreign states and representatives of other interested bodies who attended the 1992 Diplomatic Conference as observers. These observers were from Armenia, Greece, Moldova, Yugoslavia (former), the Danube Commission, UNEP, the International Maritime Organization, the World Health Organization, the Intergovernmental Oceanographic Commission, the World Meteorological Organization, and UNDP.

The Black Sea Synergy (BSS) was put forward by the European Commission (EC) in April 2007 (European Commission 2007). Three tightly interwoven factors were instrumental in the EU's decision to launch this new initiative for regional cooperation in the Black Sea area.

Through the BSS, the EU sought to increase cooperation with and between the countries surrounding the Black Sea. The Black Sea Synergy was designed as a flexible framework complementary to existing EU policies in the region, i.e. the ENP (relevant for five Eastern ENP countries: Armenia, Azerbaijan, Georgia, Moldova and Ukraine), the strategic partnership with the Russian Federation and the pre-accession policy for Turkey. Although an EU-framed initiative, the BSS was conceived as a collective endeavour, which aimed at:

- Stimulating democratic and economic reforms;
- Supporting stability and promoting development;
- Focusing on practical projects in areas of common concern;
- Responding to opportunities and challenges through coordinated action in a regional framework;
- Developing a climate more conducive to the solution of conflicts in the region.

The Black Sea Synergy was officially launched as a common endeavour at a Foreign Ministers' meeting in Kiev on 14 February 2008 which involved all regional partners, EU Member States and EU institutions (Joint Statement, 2008). One of the most important developments has been the initiative to establish sectoral partnerships in three key areas: environment, transport and energy. The Black Sea Environment Partnership has been the only one officially launched so far (notably, at a conference in Brussels on 16 March 2010). The partnerships are open to all partner countries wishing to participate, as well as to institutions that are active in the region, like the BSEC, the Black Sea Commission, the European Investment Bank (EIB) and the European Bank for Reconstruction and Development (EBRD). Reaching consensus among such a variety of actors and interests over specific projects is expected to be a cumbersome process.

The Black Sea Synergy has attracted considerable NGO interest. An alliance of 29 environmental NGOs met in Odessa on 7 February 2008 and adopted a position paper on "*Greening the Black Sea Synergy*" (June 2008).

10.2 Black Sea Economic Cooperation (BSEC)

Black Sea Economic Cooperation (BSEC), in 25 June 1992, the Heads of State and Government of eleven countries: Albania, Armenia, Azerbaijan, Bulgaria, Georgia, Greece, Moldova, Romania, Russia, Turkey and Ukraine signed in Istanbul the Summit Declaration and the Bosphorus Statement giving birth to the BSEC. It came into existence as a unique and promising model of multilateral political and economic initiative aimed at fostering interaction and harmony among the Member States, as well as to ensure peace, stability and prosperity encouraging friendly and good-neighbourly relations in the Black Sea region. With the entry into force of its Charter on 1 May 1999, BSEC acquired international legal identity and was transformed into a full-fledged regional economic organization: Organization of the

Black Sea Economic Cooperation. With the accession of Serbia in April 2004, the Organization's Member States increased to twelve.

According to the Charter of the Organization the following principles and objectives shall be promoted through the BSEC activities at various levels:

(a) to act in a spirit of friendship and good neighborliness and enhance mutual respect and confidence, dialogue and cooperation among the Member States;
(b) to further develop and diversify bilateral and multilateral cooperation on the basis of the principles and rules of international law;
(c) to act for improving the business environment and promoting individual and collective initiative of the enterprises and companies directly involved in the process of economic cooperation;
(d) to develop economic collaboration in a manner not contravening the international obligations of the Member States including those deriving from their membership to international organizations or institutions of an integrative or other nature and not preventing the promotion of their relations with third parties;
(e) to take into account the specific economic conditions and interests of the Member States involved;
(f) to further encourage the participation in the BSEC process of economic cooperation of other interested states, international economic and financial institutions as well as enterprises and companies.

BSEC covers an area of nearly 20 million km^2, a market of approximately 335 million people, and reaching an intra-BSEC trade volume of almost 300 billion US Dollars annually.

There are Five Related Bodies and Affiliated Centers:

- The Parliamentary Assembly (PABSEC)
- The Business Council (BSEC BC)
- The Black Sea Trade and Development Bank (BSTDB)
- The International Center for Black Sea Studies (ICBSS)
- The BSEC Coordination Center for the Exchange of Statistical Data and Economic Information.

10.3 EU Frameworks and Cross-Border Cooperation

The *International Commission for the Protection of the Danube River (ICPDR)* is an International Organisation consisting of 15 cooperating states and the European Union. Since its establishment in 1998, the ICPDR has grown into one of the largest and most active international bodies of river basin management expertise in Europe. The ICPDR deals not only with the Danube itself, but also with the whole Danube River Basin, which includes its tributaries and the ground water resources. The Contracting Parties to the DRPC presently include Austria, Bosnia and

Herzegovina, Bulgaria, Croatia, Czech Republic, Germany, Hungary, Moldova, Montenegro, Romania, Slovakia, Slovenia, Serbia, Ukraine and the European Union. The Contracting Parties are represented by delegations to the ICPDR, led by Heads of Delegation.

The legal basis for this international co-operation is the Danube River Protection Convention, which applies to countries with territories of more than 2000 km^2 within the Danube Basin. Apart from those the following countries also co-operate with the ICPDR under the EU Water Framework Directive: Italy, Switzerland, Poland, Albania and the Former Yugoslav Republic of Macedonia. The Contracting Parties have also committed themselves to the development of the co-ordinated international River Basin Management Plan for the Danube River Basin as requested by the EU Water Framework Directive.

The goal of the ICPDR is to implement the Danube River Protection Convention (DRPC) and make it a living tool. In addition, the ICPDR is the body that coordinates the implementation of EU Water Framework Directive and EU Floods Directive in the Danube River Basin.

The International Commission for the Protection of the Danube River (ICPDR) works to ensure the sustainable and equitable use of waters and freshwater resources in the Danube River Basin. The work of the ICPDR is based on the Danube River Protection Convention, the major legal instrument for cooperation and trans-boundary water management in the Danube River Basin.

The Danube River Basin comprises of an area of 801,463 km^2. It is the world's most international river basin, as it extends into the territories of 19 countries. The ecosystems of the Danube River Basin are highly valuable in environmental, economic, historical and social terms, but they are subject to pressures and pollution from agriculture, industry and cities—issues which are jointly addressed by the Danube Basin countries through the ICPDR.

The ICPDR acts as a platform for its contracting parties to coordinate responses to various environmental threats, formalised in the Danube Protection Convention of 1994. Since 2009, the Danube River Management Plan (DRBM) provides a roadmap for this. It contains a Joint Programme of Measures and aims to fulfil the EU Water Framework Directive (WFD). The ICPDR also implements the EU Flood Directive (EFD).

Black Sea Euroregion. The local and regional authorities of European seas have expressed for many years now, the will to develop cross-border and regional cooperation projects aimed at achieving greater prosperity, stability and security of their populations. For this reason, the congress of Local and Regional Authorities (Council of Europe) decided to launch the initiative of establishing Euroregions around these seas such as, the Adriatic Euroregion and the Black Sea Euroregion project.

These Euroregions are new associative structures, which bring together coastal cities and regions in order to achieve a sustainable management of their sea basins according to the specific needs of coastal populations. Local and regional authorities belonging to the countries part of these Euroregions can promote cooperation projects ensuring the social and economic development, the reinforcement of regional and local democracy and territorial cohesion.

The Association of Cross-Border Cooperation "*Lower Danube Euroregion*" (ACT EDJ) is a non-governmental entity, founded in April 2009, by public institutions, from Romania: Galati County Council, Braila County Council and Tulcea County Council, from Ukraine: Odessa Regional Council, Odessa Regional State Administration and Reni District Council, and from Republic of Moldova: Cahul District Council and Cantemir District Council.

The Euroregion Lower Danube comprises border areas on the Black Sea coast in the Ukraine, Moldova and Romania. Due to the historical background, the political and ethnic situation and, not least, lengthy visa procedures, cross-border co-operation in this area is particularly difficult. The Ukrainian part of the Euroregion, i.e. the area around the cosmopolitan city of Odessa with its important seaport, co-operates with the rather rural areas on the Moldavian and Romanian part of the border, which are characterised by a number of small and medium-sized towns.

The Euroregion formed several commissions whose task is to deal with the various aspects of cross-border co-operation. This allows co-operation in all areas of everyday life, including, but not limited to economic co-operation, an area where, without doubt, the most urgent problems need to be solved. In addition to this, the Euroregion addresses ecological, social and cultural issues. It tackles everyday border problems, as well as the important matter of cross-border crime.

A Strategy for the Black Sea. The EU Parliament adopted a resolution on 20 January 2011 for an EU Strategy for the Black Sea Region to be launched. Parliament stressed that the main objective for the EU and the EU Member States in this strategy is to establish an area of peace, democracy, prosperity and stability founded on respect for human rights and fundamental freedoms and providing for EU energy security. It considered that good governance, the rule of law, promotion of respect for human rights, migration management, energy, transport, the environment and economic and social development should constitute priority actions. The resolution encourages priority financing for small-scale development projects and stresses the need for a project based approach with a view to including local authorities, business communities, NGOs or other civil society organisations.

It encourages the development of synergies between the various EU policies that come into play in the Strategy, particularly the Structural Funds, the Research and Development Framework Programme and the Trans-European Transport Networks in order to ensure the sustainability of the actions financed so that opportunities created by one economic development initiative can be taken up by another, complementary initiative.

Blue Growth Strategy. The Blue Growth communication adopted in September 2012 is the maritime strand of the Europe 2020 strategy and an update of the Integrated Maritime Policy (http://ec.europa.eu/maritimeaffairs/policy/blue_growth/). This strategy consists of three components:

1. Targeted effort towards specific activities (focus areas) identified as being the most promising sectors for growth development: coastal tourism, blue energy, aquaculture, blue biotechnology and marine minerals mining.
2. Cross-cutting tools which are specific, policy integrated measures across sectors including Maritime Spatial Planning, Integrated Coastal Management, Marine Knowledge 2020 initiative and maritime surveillance.
3. Sea-basin strategies

Other aspects that are crucial for a sustainable growth in the blue economy are the development of the appropriate skills, marine and maritime research and access to finance. The EC extended the dialogue on sustainable development of the blue economy of the Black Sea to all Black Sea countries during a conference held in Bucharest, Romania on 30 January 2014.

Europe 2020 is the EU's ten-year growth and jobs strategy launched in 2010. It aims to create within the EU the conditions for economic growth:

- Smart, through more effective investments in education, research and innovation;
- Sustainable, thanks to a decisive move towards a low-carbon economy;
- Inclusive, with a strong emphasis on job creation and poverty reduction.

The Cohesion Policy is the EU's main investment tool for delivering the Europe 2020 goals within EU Member States.

The *Black Sea Basin Joint Operational Programme 2007–2013 (Black Sea JOP)* is a programme under the European Neighborhood & Partnership Instrument (ENPI) of the EU. It aims to contribute to: "*a stronger and sustainable economic and social development of the regions of the Black Sea Basin*" (www.blacksea-cbc.net).

The programme's three specific objectives, are:

- Promoting economic and social development in the border areas;
- Working together to address common challenges;
- Promoting local, people-to-people cooperation.

There are 8 participating countries in the Black Sea JOP, i.e. Armenia, Bulgaria, Georgia, Moldova, Turkey, Ukraine, Romania, and Greece.

The Black Sea JOP aims at a stronger and more sustainable economic and social development of the Black Sea Basin regions. Through the Black Sea JOP, communities in the areas concerned will be encouraged to further develop their local economy, confront environmental challenges and promote greater interaction among people.

The Programme is financed by the ENPI and its budget amounts at 25,696,516 €. The participation of Turkey is financed by the IPA (Instrument for Pre-accession Assistance). The participating countries co-finance projects with a minimum of 10% of the EU contribution.

Going further: Black Sea Basin ENI CBC 2014–2020. The participating countries' to the European Neighborhood Instrument Cross Border Cooperation Black Sea Basin (ENI CBC BSB) programme include the following:

EU Member States: Bulgaria, Greece and Romania
Partner countries: Armenia, Azerbaijan, Georgia, Republic of Moldova and Ukraine
Candidate country: Turkey
Russian Federation.

Cross-border cooperation (CBC) on the external borders of the EU is a key priority both in the European Neighbourhood Policy and in the EU's strategic partnership with Russia. CBC under the European Neighbourhood Instrument (ENI) will build on CBC under its predecessor, the European Neighbourhood and Partnership Instrument (ENPI).

ENI CBC receives funding from the European Regional Development Fund, as well as the European Neighbourhood Instrument. Both sources of funding may be used on either side of the EU external border, for actions of common benefit. The ENI Regulation sets out the basis for CBC, further defined in ENI CBC implementing rules and ENI CBC programming document.

ENI CBC aims to "*promote co-operation across the borders between EU Member States and the countries on the European Neighbourhood and Russian Federation*" and it should contribute to the overall ENI objective of progress towards "an area of shared prosperity and good neighbourliness" between EU Member States and their neighbours. Due to its geographical location, Turkey also participates in the Black Sea Basin ENI CBC as a candidate country.

Reference

Golumbeanu M, Nicolaev S (eds) (2015) Study on integrated coastal zone management. Ex Ponto Publishing House Constanta. Orginal English version, ISBN 978-606-598-397-7, 454 p. www.ec.europa.eu/maritimeaffairs/policy/blue_growth/, www.blacksea-cbc.net, www.bsec-organization.org/, www.icpdr.org/main/

Chapter 11
Conclusion

11.1 Summary

What kind of sea is the Black Sea? The Black Sea is a semi-enclosed sea component of the Mediterranean Sea (European Mediterranean or Euro-African Mediterranean) to whose main basin are linked several straits and seas: Bosphorus Strait, Marmara Sea, Dardanelles Strait and Aegean Sea.

The Black Sea cannot be a continental sea, because:

- its basin is developed both on the continental and ocean crust;
- the basin's morphology is similar to that of ocean seas (is frequently considered a miniature ocean), with continental borders and abyssal plain;
- the aquatory has active exchange relations with the Mediterranean sea and, through it, with the rest of the World Ocean.

Amongst the well-defined seas of Mediterranean Sea (Ionian Sea, Tyrrhenian Sea, Aegean Sea, Adriatic Sea, Alboran Sea, Azov Sea, Marmara Sea) the Black Sea has the largest area, the greatest depth, the largest volume and the greatest complexity.

From a geographical point of view, the Black Sea is located in the Eastern part of South-East Europe, between 40° 55′ and 46° 32′ North latitude and 27° 27′ and 41° 42′ East longitude, in a region in which the ancient stable platforms come into contract with the young mountains generated by the alpine orogeny.

The geographical location of Black Sea, as presented above, determines the main climate and biotic processes as well as the main processes of interaction between sea and atmosphere.

Latitudinal location. The fact that the middle of Black Sea is crossed by the 43° North latitude parallel, places the Black Sea in the center of the temperate zone, which has two essential implications:

E. Vespremeanu and M. Golumbeanu, *The Black Sea*, Springer Geography,
https://doi.org/10.1007/978-3-319-70855-3_11

- the presence of well marked seasons in accordance with the succession of solstices and equinoxes;
- the solar radiation ranges between 130,000 and 150,000 cal/cm^2. This annual average value of total radiation is lower than the total radiation received by the Mediterranean Sea, but sufficient to provide the energy necessary for the development of all the physical, chemical and biological processes.

The longitudinal location between 27° and 41° East longitude places the Black Sea in the way of air masses affected by the main baric centers: subtropical anticyclones of Azores, Euro-Asian continental anticyclone, North Atlantic cyclones and those from the Mediterranean Sea.

The Black Sea name (Black Sea, Mer Noire, Mare Nero, Schwarze Mer, Cernoe More, Kara Deniz) is given recently, dating since the 14th century and has no connection with the Ancient names. It seems the presence of Black Sea name in Europe is owed to Dimitire Cantemir who used and disseminated, through its scientific fame, the Latin name of *Mare Nigrum* which was translated in all the European languages.

The first information about Black Sea we owe to Ancient philosophers, starting with Homer (1000 B.C.), Hecateu (500 B.C.), Herodot (485–425 B.C.), Aristotle (384–322 B.C.), Ptolemeu (127–151 A.C.) who left texts and cartographic drawings, some of some astoundingly precise for that time. Aristotle occupies a special place within the gallery of first scholars who tried to understand the Black Sea.

Researches in the Black Sea started back in early 1800s, the first bathymetry and superficial sediments map was published in 1842. After 1850, several Russian expeditions were carried, the substantial activity of European Commission of the Danube started, the scientific investigations were developed in all the main harbors and the first marine investigations stations appeared. One of the first and most interesting syntheses on Black Sea is owed to Grigore Antipa who, in 1942, publishes the first volume of *Black Sea monography*.

The programs developed based on the investigations on board of *Atlantis II* (1969) and *Glomar Challenger* (1975) vessels resulted in the publication of various papers, including the impressive monographs *The Black Sea. Geology, Chemistry and Biology* edited under the coordination of the well-known oceanographers Egon T. Degens and David A. Ross, as well as the results of *Glomar Challenger* expedition (*Deep Sea Drilling Project*, vol. XVII).

The Knorr expedition (Black Sea Oceanographic Expedition *R/V Knorr*) conducted between April 5th to August 1st, 1988, is the first one of the series of those which deepens the knowledge on Black Sea based on a stringent measurement program using the entire arsenal of modern oceanography.

During 1991–1994 the international program CoMSBlack (Cooperative Marine Science Program for the Black Sea) is developed under the coordination of Prof. David Aubrey from Woods Hole Oceanographic Institution. The program was initiated by the executive board of Intergovernmental Oceanographic Commission) at the 25th UNESCO session, held in Paris in March 1992. This is the first major basin-wide international program, with 5 research vessels from Turkey, Ukraine and Bulgaria, properly fitted for the physical oceanographic researches and some chemical and biological features of the aquatory.

After 2000, several research programs have focused on the Black Sea and currently research of the Black Sea ecosystem is being undertaken in the framework of national, regional and international projects. The trend is towards collaborative research among riparian countries, rather than separate small-scale investigations. EU Member States (Romania and Bulgaria) comply with European Directives on marine issues (Marine Strategy Framework Directive, Water Framework Directive etc.), which required the integrated monitoring of the Black Sea parameters, in order to determine its state in relation to the target Good Environmental Status (GES).

The catchment area of the Black Sea has an area of 1,874,904 km^2 being extended in Europe and Asia Minor.

Within Black Sea catchment area four categories of catchment areas are distinguished: Western basins and North-Western basins, Crimea basins, Caucasian basins and Asia Minor basins, which transport in total 353 km^3 of water (multiannual average calculated for 1921–1988 period).

The Western and North-Western basins have the largest extension, their area being of 1,520,000 km^2 which represents 82% of the total Black Sea catchment area. In this categories falls the largest Black Sea tributaries basins: Danube, Dniester, Dnieper and Bug which transports a multiannual average liquid flow (during 1921–1988) of 261 km^3/year, representing 76% of the flow tributary to Black Sea of its entire catchment area, with variations between a minimum of 153 km^3, in 1921, and 389 km^3, in 1970. The solid flow represented by suspended alluvia, dissolved salts, mineral nutrients and organic matter, is of roughly 150 million tonnes per year, having an essential role in the operation of aquatory system and in the distribution of recent sediments.

The Crimea basins have a total area of only 2729 km^2 representing 0.14% of the total catchment area. The largest rivers are Cernaia, Belbek, Alma and Kacea which transport a multiannual average (1921–1988) of 0.32 km^3 of water (0.09 of the total), with variations between a minimum of 0.07 km^3 in 1930, and a maximum of 0.67 km^3 in 1968.

The Caucasus basins have an aggregate area of 75,000 km^2 representing 4% of the total. The most important rivers are: Cuban, Rioni, Ciorok, Inguri, Kodori, Bzib, Supsa and Mzimta. The average of multiannual liquid flow is of 43 km^3 (13% of the total) with variations between a minimum of 31 km^3, in 1969, and a maximum of 57 km^3 in 1922.

The Minor Asia basins have an aggregate area of 259,550 km^2 representing 14% of the total catchment area. The most important rivers are Yeşilirmak, Kâzâlirmak and Sakarya. The multiannual average flow, for the period 1930–1986, is of 36 km^3 (11% of the total) with variations between 25 km^3, in 1949, and 51 km^3, in 1940.

Danube basin and Danube Delta. The Danube basin (817,000 km^2) represents 43.57% of the total area of Black Sea basin having an essential influence over its Western compartment. The river's place of birth is considered the confluence point of two streams, Brigach and Brege, near Donaueschingen locality from

Baden-Wurtenberg region, Germany. Here is the 2863 km which represents the route that the Danube shall go through until Black Sea, finally reaching to a basin area of 817,000 km^2. On this surface the farming lands prevail (44%) and meadows of various types and uses (30%), the forests representing only 26%. The geographic criteria allow the Danube basin to be divided into three sectors: Upper Danube, Middle Danube and Lower Danube. *The floods regime* was, in natural regime, essential for the operation of Danube Delta and Floodplain. After several and various developments which led to the pronounced artificiality of hydrological regime, the floods' role decreased a lot. According to the amplitude and duration, we differentiate floods with small amplitude and duration, with moderate amplitude and small duration, with moderate amplitude and moderate duration, with high amplitude and long duration, with very high amplitude and long duration (Vespremeanu 1993). The construction of Hydroenergetic and Navigation System from Iron Gates I is located between Gura Vaii and Sip and the reservoir formation within Danube Gorge is the most severe intervention having the most numerous and intense consequences for the artificiality of flow on Lower Danube. The flow's artificiality involves also the modification of solid flows regime but to a lower extent and with less important consequences. We mention two directions of loss of liquid flows. The first one is connected to the completion of Main-Danube channel. Its length between Regensburg and Bamberg is of 204 km. Here, a flow of roughly 466 million m^3/year runs from the Danube's basin in Main basin. The second direction, with lower losses of flows, is due to irrigations. Only Bulgaria takes out from the Danube over 150,000 million m^3 of water per year for irrigation.

The consequences of flow's artificiality, especially of solid flow, are felt in full force not only by Danube Delta but also by the deltaic and lagoon coast as we will see further. In 2000, the population of Danube's basin was of 77,722,460 inhabitants which represent a density of 186 inhabitants/km^2. This impressive density together with the industrial and agricultural activities, contributes largely to the pollution with toxic substances, to the microbial pollution and to the release in Black Sea of some large quantities of nutrients which have a catastrophic contribution to waters' eutrophication.

The Danube Delta is famous as one of the greatest wetlands of the earth. The wonderful natural habitats formed here offer good living conditions for an impressive number of plants and animals. Among these, reeds form one of the largest single expanses in the world, and Letea and Caraorman forests represent the northern limit for two rare species of oak that are more frequently met in the south of the Italian and Balkan peninsulas. Together with the great number of aquatic and terrestrial plants, there are also many important colonies of pelicans and cormorants, which are characteristic of the Danube Delta, as well as a variety of other water birds which reside in or visit the delta for breeding or wintering. The large number of fish is also notable, with species of both high economic and ecological value (Golumbeanu and Nicolaev 2015).

Geology and geophysics. The Black Sea basin (called also Euxinic basin) is located in the Southern part of lithospheric plate near the contact with African and Arabic lithospheric plates.

The Black Sea basin is a tectonic basin consisting of two compartments, Western and Eastern, separated by Andrusov crest with a horst appearance. The entire region overlaps a basis developed on oceanic crust and continental crust unevenly distributed in West-East transversal profile of the basin.

The typical oceanic crust is on the central part of the Western basin, under the Euxinic Abyssal Plain delimited by the isobath of −2000 m. Here, the level of Moho discontinuity rises to the surface, reaching in the central part of the basin at only 20 km below the actual level of the sea. The basaltic layer, with a maximum thickness of 8 km, overlaps the upper mantle which has temperatures of 500–600 °C and is covered by non-consolidated sediments with a 7–14 km thickness.

The continental crust is located on the borders of the Western basin and is characterized by the descending of Moho discontinuity until roughly 35 km depth, by the thickening of basaltic layer which reaches 15–18 km and by the wide extension of the granitic layer covered by other types of consolidated sediments. The Eastern basin is overlapped to a continental crust with basaltic layer of 8–9 km thickness over which is laid a thin granitic layer covered by non consolidated sediments developed on a 10–12 km thickness. This continental crust is increasingly thickening towards East, where it reaches thicknesses of over 40 km.

In the central part of the Black Sea the Andrusov Crest is located made of continental crust with thicker basaltic layer over which a thin granitic layer is laid and then the pile of non-consolidated sediments. At East of Andrusov Crest we cannot speak about typical oceanic crust due to the presence of thin granitic layer which covers the basaltic layer.

The *relief* of the Black Sea basin is divided, in terms of morphotectonic and morphostructure, in two regions: central and boundary.

The central region, developed on the oceanic type basaltic crust and partially on the continental crust, encompasses the Euxinic Abyssal Plain located at depths under 2000 m. The primary tectonic relief was buried by the sediments deposited during the last 80–100 million years so that the actual relief, developed on fine and ultra-fine non-consolidated sediments is particularly plain and smooth.

The straits that link Black Sea to other neighboring seas are Bosphorus Strait and Kerch Strait.

Bosphorus Strait. The Black Sea is connected to the Mediterranean Sea and, through it, to the World Ocean by the Turkish Straits represented by Bosphorus Strait (also called Istanbul Strait) and Dardanelles Strait (also called Canakkale Strait), which ensure an active exchange of waters, having a strong influence over the entire structure and operation of aquatory. Bosphorus Strait is located between Istanbul Peninsula (Europe) and Kocaeli Peninsula (Asia Minor), crossing a region with a very complex geological and tectonic structure. *The strait's length* is of 31 km. The maximum depth is of 110 m, and the minimum depth of 33 m, the average depth 35.8 m. Within the strait there are two salients, one in South, near Beşiktaş, at 33 m depth and another in North at −61 m. The strait's length ranges between a minimum of 0.7 km and a maximum of 3.5 km. The average multiannual water flow through Bosphorus is the following:

- 600 km^3/year or 20,000 m^3/s of mean salinity waters of 17.90% flow from Black Sea to Marmara Sea on surface
- 300 km^3/year (10,000 m^3/s) of mean salinity waters of 35.50% flow from Marmara Sea to Black Sea on the bottom.

The water flows values through Bosphorus Strait, as well as their stratification varies a lot during the year according to the levels from Black Sea and the prevailing winds. The entry in Black Sea of an average quantity of 300 km^3/year, with an impressive flow of 10,000 m^3/s of Mediterranean origin water more salty and warmer than the own waters leads to, by multiples diffusions, essential changes of Black Sea aquatory structure. **The Strait of Kerch** is less important than Bosphorus Strait, due to the small sizes of Azov Sea but especially to the drastic reduction of river input following the hydrotechnical development of Don River, completed in 1953.

The Strait of Kerch links Black Sea to Azov Sea, having a length of 45 km and a width ranging between 3.7 and 52 km. The depth varies a lot in longitudinal profile: 18 m in South, 5 m in the central part and 10 m in North. The Strait of Kerch consists of a succession of four narrow sectors, including Taman, Kerch and Kamâş Burun Bays.

Through the Strait of Kerch an opposite circulation takes place which ensures waters exchange with Black Sea. Therefore, from Azov Sea, between 22 and 95 km^3 of 8–10% salinity waters flows into the Black Sea every year, on surface, and from the Black Sea between 29 and 70 km^3 of 16–18% salinity waters enters annually in Azov Sea at depths of 5–7 m. *The Azov Sea* is actually a brackish firth adjoining Black Sea (the Don Firth would be a more appropriate name) linked by the Strait of Kerch. Its shape is elongated, with the big axis oriented North-East towards South-West. The maximum length is of 400 km, *maximum width* is of 200 km, and *average width* of 95 km. The Azov Sea area is of 38,800 km^2.

The knowledge of the Black Sea evolution requires the analysis of two issues: the morphogenetic evolution whereby the basin reached its present day basin, briefly described above, and the evolution of the aquatory, i.e. the sequence of events which culminated in the current model of water mass in the basin. The issue of the evolution of Black Sea aquatory from late Neogene, until present day, was subject to various studies and reenactments based on the analysis and age dating of transgressive sedimentary formations.

Oscillations of the Black Sea level. It is obvious that we are witnessing rises of the Black Sea level with rhythms between 0.83 and 5.66 mm/year, according to the various compartments of the coast. The very high differences may be explained only by the different isostatic behavior of continental margins. Such processes took place also in the past, giving an explanation for the emergent and submergent oscillations with quite large amplitude of the sea level obviously indicated by archaeological data.

The surface waters temperature, as well as the air temperature are characterized by significant differences between the two compartments, Western and Eastern, between which the central part with transition character is located. The largest differences are recorded during winter and those during summer are insignificant. During autumn, the cooling of surface waters starts from North-West (Fidonisi Bay) where in September the average temperature is of 18.4 °C, towards the South-Eastern corner, where the average temperature in September is of 21.2 °C.

The average annual surface salinity ranges between a minimum of 14.5%, in the high seas of Danube Delta, and a maximum of 20.7% in the central part of the basin. In front of Danube's mouths, the salinity drops to 5–8% due to the large volume of river water. The average annual surface salinity has a longitudinal distribution, with low values and high gradients on the Western and Eastern sides and with values exceeding 18.20% in the central part of the basin.

The vertical distribution of density presents the following levels:

- Superficial layer which lies between the surface and −10 up to −15 m, with sigma-t densities of 10.5–11 kg/m^3;
- Permanent pycnocline layer which lies between −15 and −120 m where the sigma-t density rises from 11–11.5 to 15.5–16 kg/m^3, with a gradient of 5–6 kg/m^3;
- Deep layer which lies at depths greater than 120 m, with sigma-t density of 16–16.5 kg/m^3.

We mention the importance of firm pycnocline presence with high gradient which has a decisive influence on the development of vertical transfer processes. However, the intense wind activity during winter generates frequently convection processes which determine the local ventilation of pycnocline and of deep waters up to depths of 400–500 m.

The dissolved oxygen. The amount of dissolved oxygen registers maximum values in the superficial layer between surface and −10 m where it ranges between 330 and 380 μM/l (micromoles per litre).

At 100 m depth, the oxygen quantity drops until below 5 μM/l in the central part of the deep basin, maintaining however at values of 275–300 μM/l in the close proximity of continental slopes. Below 150 m, the oxygen drops fast and disappears between −190 and −200 m.

The hydrogen sulphide and sulphides in the Black Sea water. The vertical distribution of the average quantity of H_2S illustrates a continuous rise from 150 m depth up to −2000 m.

The analysis of production distribution and H_2S consumption showed the existence of a horizon with maximum production between 500 and 1000 m depth, of a maximum consumption horizon higher than −500 m and a neutral horizon below −1000 m.

The vertical position of anoxic-suboxic interface is very unevenly distributed, resulting in a "relief" with "hills" and "depression" which varies in time.

Water masses. In the Black Sea there are three water layers with original properties specific only for this sea. These are the cold intermediate layer (CIL), suboxic layer (SOL) and anoxic layer. To these are added the Mediterranean water masses which enter through Bosphorus Strait in the Black Sea.

The Cold Intermediate Layer (CIL) is characterized by a thermal inversion, being bordered at the upper and lower part by the hydroisotherm of 8 °C. CIL encompasses waters with minimum temperature of 7.10–7.20 °C and salinity of 17.5–18.2%.

The Suboxic Layer (or SOL), was recently identified and studied. SOL is individualized as a very stable layer between the lower part of the oxycline, with an upper limit at 55–60 m depth and a lower limit at −120 m, rarely to −180 m. SOL is the layer where the oxygen quantity drops, without disappearing and the H_2S quantity rises. **The anoxic layer** lies below −200 m and has a water volume of about 500,000 km^3, until recently considered a stable and inert water mass, with no dynamic activity.

The currents from the Black Sea. Recent researches proved the existence of several circulation types in the Black Sea basin: the major surface circulation from the deep basin, the sub-basin and mesoscale surface circulation, upwelling and downwelling vertical circulation.

Major surface circulation in the deep basin. The model accepted now, which we may call the "*Oguz model*" illustrates the presence in the deep basin of a cyclonal current called Black Sea Rim Current. This current flows actively in the layer between the surface and −150 m, has a geostrophic character and speeds of 15–20 cm/s at surface and over 20 cm/s in the core. The current speed drops as the depth is greater, being felt up to −500 m, where a slow circulation exists, with maximum speeds of 2 cm/s, very variable, with countercurrents and vortexes (eddy) to which the recycling cells are associated. The average total water transport is of 5.75 million m^3/s (5.75 Sv), with important seasonal variations. During winter, the total transport of water is of 6 Sv, spring—8 Sv, summer—4 Sv, and autumn—5 Sv (1 Sv = 1 million m^3/s).

The Rim Current has 50 km width, doubles slightly with wave width of 150–200 km and flows at the deep basin's border, beyond the shelf edge, over the continental slopes. In transversal section shearing processes occur frequently, which determines the formation, towards the exterior, of 9 well shaped anticyclonic vortexes called Sevastopol, Kaliakra, Bosphorus, Sakarya, Sinop, Kazalirmak, Batumi, Caucasus and Crimea, to which are added several small, secondary vortexes. The Rim Current separates the coastal waters with surface salinity of 15.44–17.97% from the deep basin waters with surface salinity of 18.54–19.00%.

The sediments on the bottom of the Black Sea deep basin are grouped in three lithological units deposited in the last 25,000 years:

- The 30 cm thick unit 1 consists of banks with coccolithophores rich in carbonates (60–80% $CaCO_3$) and low organic carbon (5–10%) deposited in the last 3000 years.
- The 40 cm thick unit 2 consists of micro-laminated sapropel deposited during 7000–3000 B.P. Starting with 7090 + 180 years B.P. the quantity of organic carbon from sediments rises sharply to over 80% while the quantity of carbonates maintains itself at values of 15–20%.

- Unit 3 has a not-specified thickness, continuing itself below the core level and consists of black laminated lutite alternating with coarser fractions, rich in iron sulfates and low in organic carbon (under 5%) and carbonates (4–5%).

The interface between Unit 2 and Unit 1 represents the fast shift from lacustrine regime existing in the Black Sea about 7200 years ago to the marine regime after the invasion of Mediterranean waters through Bosphorus Strait.

The present day sediments are of terrigenous, biogenous and autigenous origin.

The terrigenous sediments are transported by the Black Sea streams and rivers, as well as by the strong winds which carry large quantities of dust from the pontic steppes. The smaller quantities of terrigenous sediments come from the abrasion of loess shores.

The river terrigenous sediments represent the most important source and the total quantity of detritic material transported from the Black Sea amounts to 149.45 million t/year. The solid flows transported in suspension are distributed vey unevenly: 83 million tonnes (55%) are transported by the Danube, 37 million tonnes (25%) come from the Caucasus Mountains, 11 million tonnes from the North pontic plateaus and plains and 17 million tonnes (11%) from the Pontic Mountains and Anatolia Plateau. Most of these sediments are transported in suspension during spring, except for the sediments transported by the Caucasus streams, which register maximum solid flows during summer.

The streams and river also transport large quantities of ions found in solution. The annual chemical input flow in the Black Sea is of 80 million tonnes, of which 65% is attributed to the Danube. To this are added the mineral nutrients (nitrogen, ammonia, phosphates, silicones) with a total of 1.2 million tonnes.

The biogenous sediments come especially from the phytoplankton. The production of organic matter is of 812 million tonnes/year, which represents a lot comparing to other seas. The high values of the primary production are explained by the important input of nutrients, by the chemical exchanges between the anoxic layer and suboxic layer with the oxic layer, as well as through the low levels of salinity which favor the development of phytoplankton. The main groups present in the phytoplankton are the diatoms, dinoflagellates and coccolithophores, all reaching the maximum biomass during summer. Under these conditions, the average quantity of organic carbon is of 3 mg/l, twice larger than the average quantity from the World Ocean.

The autigenous sediments result from the chemical exchange processes developed in the biogenous sediments, below −1800 m, on the abyssal plain. Here, the pyrite type colloidal and meha-colloidal iron disulphides (Fe_2S) are continuously formed through the decomposition of the organic remaining under anaerobic conditions.

The distribution of sediments is influenced by the main dispersion factors: waves and currents. *In the coastal areas* (littoral and continental shelf), the waves and waves-generated currents, as well as the wind coastal currents are the main factors providing the dispersion of sediments. The littoral sediments are dominated by the terrigenous, terrigenous-biogenous or biogenous-terrigenous sands, according to the predominance of such fractions. The *Mya, Mytilus, Venus, Cardium* shells have a 10–60% contribution to the composition of shore and beach sand. The important input of rivers is stored mostly in deltas, lagoon and littoral fronts.

The continental shelf sediments differ a lot from one region to another. On the Danubian shelf the shell carbonate sediments prevail, with over 50% $CaCO_3$, mixed with clay minerals. On Anatolian and Scythian shelves there are banks with biogenous calcite mixed with biogenous aragonite and terrigenous carbonates. The Caucasian shelves are characterized by the presence of biogenous coccolith clay muds mixed with terrigenous calcite. The sediments on the continental slopes are differentiated according to the type and position of such slope.

On the Moessic and Scythian *continental slopes* terrigenous carbonate muds (10–15%) and mitilloide and phaseolonic muds are deposited. On the Caucasian continental slopes the clay muds are prevailing, to which are added the terrigenous muds low in carbonates. The Pontic continental slopes are covered by a large variety of sediments, with the predominance of the terrigenous muds lows in carbonates mixed with coarse sediments.

The continental glacis sediments are represented by muds and grey clays low in carbonates and rich in clay minerals, as well as by greenish-grey organic-terrigenous muds.

On the abyssal plain there is an alternation of smooth coccolitic muds and greenish-grey silty clays rich in organic matter, iron sulphides and sapropel.

Flora and fauna of the Black Sea. *The plants* are represented by over 304 species of macrophyte seaweeds, most of them red seaweeds, to which are added the brown and green seaweeds. The phanerogams are represented by 5 species only.

The animals are represented by the majority of invertebrates groups, with a total of 1750 species, and the vertebrates include fishes, birds and marine mammals, with total of 164 species.

The associations of marine organisms are: plankton, nekton and benthos. *The plankton* consists of the total organisms which live by passive floating in the water mass, grouped in two categories: phytoplankton and zooplankton.

The phytoplankton is formed mostly of diatoms (136 species, representing about 80% of the phytoplankton composition), followed by the dinoflagellates (about 17%), cyanophyceae, chlorophyceae, silicoflagellates, in total 269 species of seaweeds. The most frequent species of diatoms belong to the *Chaetouras* and *Coscinoides* genera.

The zooplankton is formed of 70 species, most of them tintinnids, rotifers and copepods, which prevail, as well as of cladocera, jellyfish and chaetognatha. The number of species forming the zooplankton is surprisingly low comparing to that from the Mediterranean Sea. Of particular interest is the stenotherm plankton adapted to the life in cold waters, which during winter and spring is present in the superficial waters, and during summer descends in the cold intermediate layer (CIL). The nekton is formed of planktofag fishes such as the anchovy (*Engrauluis encrasicholus*), small horse mackerel (*Trachurus trachurus*), mullet (*Mugil sp.*), Atlantic mackerel (*Scomber scombrus*), and predatory fish like the Atlantic bonito

(*Sarda sarda*), Mediterranean horse mackerel (*Trachurus mediterraneus*) and benthofags, like snouted sole (*Solea nasuta*), flounder (*Pleuronectes flesus*), turbot (*Scophthalmus maeoticus*), sturgeons in the early stages.

The benthos is formed of 1790 species, most of them including polychaeta, nematodes, molluscs, crustaceans, bryozoas, echinoderms, tunicates, forming various biocoenosis up to the depth of almost 200 m.

Fishes, birds and mammals from the Black Sea. The fishes are represented by three species from the *Chondrichthyes* class and 164 species from the *Osteichthyes* class.

The Chondrichthyes class is represented by the piked dogfish (*Squalus acanthias*), thornback ray (*Raja clavata*) and the common stingray (*Dasyatis pastinaca*). The dogfish is widespread in the shelf waters and the deep basin waters, at depths of 20–80 m, and frequently comes close to the shore.

The most popular, by their economic value, are the anadromous migratory species: Atlantic mackerel, Atlantic bonito, horse mackerel, bluefish, anchovy and mullet. In the high seas are present the tuna (*Thunnus thynnus*) and swordfish (*Xifias gladius*).

The gobies, bleniidae, labridae, shallow-waters fish live in the regions with rocky substrate, and the snouted sole, flounder, turbot, big-scale sand smelt, sturgeons live in the regions with sandy substrate.

The sturgeons (*Acipenseriformes* order) are fish with bone cartilaginous skeleton, present in the Black Sea coastal waters, from where they enter the rivers during the spawning period. The four species have a particular economic value, both for their tasty meat as well as for the caviar sought after in the fish markets: beluga (*Huso huso*), starry sturgeon (*Acipenser stellatus*), sterlet (*Acipenser rhutenus*) and Russian sturgeon (*Acipenser gueldenstaedtii*). In present, the populations of these species are increasingly reduced in terms of quantity due to the overfishing and the degradation of benthal biocoenosis where they are feeding.

A fish that appears temporary in the Black Sea is the European eel (*Anguilla anguilla*), which reproduces in the Sargasso Sea of Atlantic Ocean. The larvae and fry are transported by the Bay's Currents and the North Atlantic Drift for 2–3 years, reaching the high seas of Gibraltar Strait. Young eels (cives) pass, through Gibraltar Strait, in the Mediterranean Sea, entering on the streams flowing in it. Part of them arrives through the Aegean Sea and Marmara Sea in the Black Sea where they stay for a while in the littoral waters. From here they enter in Danube up to the upper basin, including on Romania streams, where they stay for 10–15 years, until sexual maturity, when they try to restore the way back to Sargasso Sea for spawning. Most of them die in the Mediterranean Sea and in the Atlantic waters in front of Gibraltar Strait. The population shall be restored by the eels that migrated on the American streams and come back, under proper conditions, in Sargasso Sea.

The typical marine ***birds*** are few and appear accidentally in the Black Sea space. This is the case of the manx shearwater (*Puffinus puffinus*), noticed above the littoral waters and in high seas especially during pre or post-nuptial migrations. The marine shore birds are much numerous, but they have a very large area of activity in the inland waters, often at large distances from the shore.

The mammals are represented by two species of dolphins, seal and porpoise: common dolphin (*Delphinus delphis ponticus*), common bottlenose dolphin (*Tursiops truncatus ponticus*), monk seal (*Monachus monachus*) and harbor porpoise (*Phocoena pho-coena*).

The Black Sea fauna was formed during the basin's evolution, starting with the Sarmatian Sea until now. Therefore, in the actual fauna we find Sarmatian relict, Mediterranean species, Atlantic species, Indo-Pacific species and freshwater species. From the total of 1650 species of invertebrates and vertebrates, most of them (1320 species) have Mediterranean origin, 305 species are freshwater and only 125 species are relict.

Sarmatian relict. Within the wide Sarmatian basin which was shared from the Vienna basin, Pannonic basin, Getic basin, Euxinic basin, over the actual Black Sea plain, through Manici sleeve, until the Caspian and Aral basins, a brackish fauna existed well conserved in the Caspian Sea, where it was protected against other marine influences. This fauna is present in the Black Sea as Sarmatian relict called also Ponto-Caspian relict. Amongst the invertebrates, this category includes gastropods, polychaeta, isopods, amphipods, mysis shrimp species. Amongst the vertebrates, the fishes are most numerous: most of sturgeon, mackerel and gobiidae species.

The Mediterranean origin fauna entered the Black Sea during at least three stages, the last one starting 7200 years ago. This category includes the majority of invertebrates and vertebrates species populating today the Black Sea waters.

The freshwater fauna includes euribionte species such as *Percaria demidoffi*, *Acipenser nudiventris* and *Acipenser gueldenstaedtii*, which do fine at quite large variations of the salinity.

Recent invaders. Through various ways, but especially through the vessels' hulls and in the ballast waters of the oil tankers, in the last 100 years 21 species of animals were accidentally introduced. From the Atlantic Ocean come 7 species, of which 2 are crustacean cirripede (*Balanus improvisus* and *B. eburneus*), 2 are jellyfish (*Blackfordia virginica* and *Bougainvilia megos*), 2 are molluscs (*Mya arenaria* and *Doridela obscura*) and a species of ctenophora (*Mnemiopsis leidy*). From the Pacific Ocean come the *Rapana thomasiana* gasteropod and *Cunearca cornea* bivalve. From the North Sea come 4 species, of which 2 are crabs species. Most of these invading species accommodated well in the Black Sea waters, competing the local species and generating ecological imbalances.

Deliberately introduced species. For economic, medical and aquarium related purposes, 13 species were deliberately introduced in the Black Sea. For economic purposes, 2 species of crustacean were introduced, 1 species of bivalve and 9 species of fish. For medical purposes, it was introduced, from Italy the *Gambusia affinis* fish for fighting against the malaria transmitting mosquito, knowing that gambusia is a great consumer of mosquito larvae. For aquarium purposes, the *Oryzios latipes* was introduced from Japan.

The coastal environment is a system consisting of several sub-systems closely interconnected: relief, sediments, aquatory, atmosphere, ecosystems and social and economic subsystem. This very dynamic, fragile and vulnerable system lies over a water line length of over 500,000 km and an area of roughly 50,000,000 km^2 of the Earth area. Here are the large urban agglomerations of the planet and two thirds of the World population lives here. Therefore, the human pressure on this environment is very intense, manifesting itself by excessive pollution, extension of hyper-built spaces and destruction of ecosystems.

In the last 30 years it is noticed the intensification of coordinated attack of natural risk factors over the coastal environment on the whole and especially over the shore, which is the most dynamic and vulnerable part. The first attack comes from the sea and is due to the rising of waters level with rhythms ranging between +1.5 and +5 mm/year. The second attack is due to the increase in severe storms numbers. Both natural hazards are the results of the imbalances installed at the atmosphere-sea level interface due to global warming and pollution.

Following the human pressures and the actions of the natural hazards, the coastal environment is now subject to degradations which tend to become irreversible. Such degradations affect especially the ecosystems and the shore's relief. Annually, millions of hectares from the World Ocean shore are damaged or transformed, which represents priceless losses for the riparian states. Not only the economy is affected but also the population, human settlements, harbors and health resorts.

All these issues, as many others, imposed the development of concerns for the **coastal zone management**, which aims: "*solving the present and future issues from the coastal zone for ensuring a sustainable balance between the economic uses and a better quality of the environment. This aim may be achieved by using the deepened analysis of the natural processes and of the social and economic development*".

Human pressure over the Black Sea. The activities of population bordering the Black Sea induce a huge anthropogenic pressure. Considering the entire surface of the Black Sea catchment area, the pressure is, of course, larger. This pressure is exercised by pollution, eutrophication, overfishing, endangerment of rare species, loss of biodiversity, introduction of new species, alteration of structure and operation of wetlands surrounding the sea, shores degradations, alteration of landscape aesthetic.

The population and human settlements from the coastal regions. The states bordering the Black Sea are Romania, Ukraine, Russian Federation, Georgia, Turkey and Bulgaria which in 2004 had a total of 294,316,181 inhabitants. He states bordering the Black Sea are Romania, Ukraine, Russian Federation, Georgia, Turkey and Bulgaria. The Black Sea regions have a permanent population of roughly 17.5 million inhabitants, to which are added 6–8 million tourists per year. This population is very unevenly distributed in the riparian states. The Black Sea catchment area has registered a total number of 63 harbors, as follows: Romania: 18 harbors (including rivers), Ukraine: 18, Bulgaria: 2, Moldova: 1, Turkey: 24 (Black and Marmara Sea). The important harbors on the Black Sea coast are: Constanta and Sulina (Romania), Odessa, Mariupol (Ukraine), Tuapse, Poti and Batumi (Georgia), Samsun and Istanbul (Turkey), Burgas and Varna (Bulgaria).

In general tourism along the Black Sea coastal zones is rather well developed in Bulgaria, Romania and Ukraine. Besides Istanbul, Odessa and Varna are the largest cities within the Black Sea region, with a high capacity to attract tourists.

International cooperation in the Black Sea Basin. Strategically placed between the European Union, candidate countries and neighbors, the Black Sea region constitutes an axis of increasing geopolitical importance for Europe. With its vast marine, energy and other natural resources, unique ecosystems and diverse interests ranging from shipping to tourism and fisheries to oil pipelines, nowhere is regional cooperation busier or more diverse. The governments of Black Sea countries are signatory parties to several international treaties relating to regional environmental cooperation. The significant role of the national governments in integrated approach of coastal zone should be realized through cooperation and partnership with other governments, the business sector and local communities. Today, the *Convention on the Protection of the Black Sea Against Pollution* (also referred to as *Bucharest Convention*) is one of the most known European Regional Sea Conventions, establishing the legal ground for combating pollution from land-based sources and maritime transport, achieving sustainable management of marine living resources and pursuing sustainable human development in the Black Sea Region.

11.2 The Importance of the Black Sea and Its Preservation

A considerable part of Europe's natural heritage surrounds the Black Sea. Much of the Black Sea Region (WWF 2006) hosts species and habitats that have disappeared elsewhere, or that are unique to these areas. But the region faces major environmental challenges: some of them inherited from the past or due to the hardship that most of the region's countries underwent after the collapse of the communist regimes, but some may result, or already result, from new developments. These different environmental challenges call for different forms of international response.

Where environmental pressures affect a collective resource, such as fisheries or a shared watercourse, international co- operation should focus on its shared management: this applies to the Danube river and its watershed, or to nutrient pollution and fisheries in the Black Sea.

But the region also remains economically underdeveloped, and poverty, poor administrative capacity and inadequate legislation constitute major problems as they lead to the inadequate management of resources exemplified by poaching, illegal logging and overgrazing. These pressures affect, first and foremost, local resources, although their cumulative effect may affect resources on a national or regional level. Improving capacity and governance locally and nationally is key to tackling them. International cooperation can contribute significantly to building capacity through the exchange of best practices and through assistance. International support is of course already being provided and should continue to be available in this field. In the years to come the development of domestic capacity will depend primarily on the determination and capability of each of the region's governments (WWF 2008).

In addition, the economies of all countries in the region are growing, some of them quite fast. Industrial and urban development and infrastructure projects, stalled in the 1990s, have resumed and will increasingly add to environmental pressures (EEA 2015). This is of particular concern where those promoting such projects have access to vast resources and influence, out of proportion with those available to local administration and conservation authorities. A strategy for sustainable development in the Black Sea Region must anticipate these developments too (WWF 2008).

11.3 Recommendations

- **COOPERATION**: Cooperation between Black Sea countries and the EU should focus in particular on monitoring and reducing the environmental impact of large infrastructure developments, including oil, gas, and transport infrastructure. In fact, environmental impact assessment is subject to European legislation in the Black Sea Member States (Romania and Bulgaria)—SEA Directive 2001.
- **FUNDING**: promotion of environmental protection and conservation policies on the basis of EU legislation. Cooperation with non-EU member countries should address conservation priorities through the new emerging funding opportunities, such as Horizon 2020, ERA NET, Joint Operational Programme Black Sea Basin 2014–2020 etc., as well as national funding programmes. National governments should be encouraged to allocate appropriate resources to nature protection and authorities responsible for nature protection should explore, with international support, alternative financing systems to supplement public funding when it is deficient; strong scientific research capability is essential to sustain the protection of biodiversity and this should be supported by Black Sea governments and by the EU.
- **AWARENESS**: more awareness-raising and greater emphasis should be placed on the practical consequences of environmental degradation and on the benefits of conservation. Environmental advocacy and public participation should be founded on local concerns and values. Non-governmental organizations and the media have an essential role in building public support for environmental protection and they have contributed much over the last decade in this respect and their efforts must be supported, aiming at the protection of the Black Sea.

References

Directive 2001/42/EC on the assessment of the effects of certain plans and programmes on the environment. Available online at: http://ec.europa.eu/environment/eia/sea-support.htm

EEA (2015) State of Europe's seas, EEA Report No 2/2015. European Environment Agency, Copenhagen, Denmark

Golumbeanu M, Nicolaev S (eds) (2015) Study on integrated coastal zone management. Ex Ponto Publishing House Constanta. Orginal English version, ISBN 978-606-598-397-7, 454 p

Vespremeanu E (1993) Oceanografie. Ed Univ. Bucureşti, p 410

WWF and Heinrich Böll Foundation EU regional office Brussels (2008) Greening the Black Sea Synergy

WWF/Williams L (eds) (2006) An ecoregional conservation plan for the caucasus. WWF. Tbilissi, May 2006. Available online at: http://assets.panda.org/downloads/ecpsec-ond_edition.pdf

MIX
Papier aus verantwortungsvollen Quellen
Paper from responsible sources
FSC® C105338

If you have any concerns about our products,
you can contact us on
ProductSafety@springernature.com

In case Publisher is established outside the EU,
the EU authorized representative is:
Springer Nature Customer Service Center GmbH
Europaplatz 3, 69115 Heidelberg, Germany

Printed by Libri Plureos GmbH
in Hamburg, Germany